DIFFUSION PROCESSES, STRUCTURE, AND PROPERTIES OF METALS

PROTSESSY DIFFUZII, STRUKTURA I SVOISTVA METALLOV

ПРОЦЕССЫ ДИФФУЗИИ, СТРУКТУРА И СВОЙСТВА МЕТАЛЛОВ

Diffusion Processes, Structure, and Properties of Metals

Edited by
S. Z. Bokshtein

Authorized translation from the Russian

Springer Science+Business Media, B.V.
1965

ISBN 978-1-4899-4958-5   ISBN 978-1-4899-4956-1 (eBook)
DOI 10.1007/978-1-4899-4956-1

This collection of reports was originally published by the
Mechanical Engineering Press, Mashinostroenie, in Moscow in 1964.

Library of Congress Catalog Card Number 65-10718

CONTENTS

PREFACE

The relationship between the structure and the properties of metals depends on the elementary processes occurring in metal alloys and the elementary processes which destroy the crystal lattice. These two types of processes are interrelated, although the relationship is not yet clear.

Diffusion is responsible for structural changes in metals, but is itself dependent on the structure of these metals. Separation surfaces such as grain and block boundaries, boundaries between phases, and the surface layers of metals have a considerable effect on the diffusion process. Structural changes occurring as the result of phase transformations or recrystallization processes also affect the kinetics of diffusion processes. The rate of diffusional displacement of atoms depends not only on structural factors but also on external factors—fields of force, in particular—which induce stress and plastic deformation in metals. The structural singularities of metals and the actual operating conditions have not been adequately taken into account in the numerous experimental and theoretical investigations made up to the present time.

Many machine parts operate under conditions of very high temperatures and complex stresses—conditions in which diffusion becomes a very important problem. Also, the high-melting-point metals which are now being used in industry (molybdenum, niobium, and tungsten, for example) require special methods of investigation.

Modern machines require very strong materials, and it is the necessity of manufacturing such materials which impels scientific studies. The problems of very high strength are being solved, first, by the preparation of crystals with a small number of defects, the so-called threadlike crystals (whiskers), and second, by the preparation of crystals with a large number of defects with low mobility by creating a very fine, uniform, highly dispersed structure characterized by a high density of dislocations with low mobility. For this latter purpose a combined treatment is used, thermomechanical treatment in particular.

All these problems are treated in this book. The book is divided into four parts: diffusion and the structure of metals; recrystallization; pore formation and destruction at high temperatures; and high strength resulting from the preparation of threadlike crystals and from thermomechanical treatment.

Radioactive isotopes were used in almost all these studies; it is a very effective method of studying localized processes.

SOVIET ALLOY DESIGNATIONS

The alloy designations in this volume have been transliterated from the Russian rather than translated to avoid confusion with Western alloys that are similar but not identical. The explanation of these designations that follows is based on the one contained in the Handbook of Soviet Alloy Compositions (U.S. Department of Commerce, OTS, Washington 25, D.C., PB 171331) and the British Iron and Steel Industry Translation Service Publication No. BISI 2000. A complete description of steel and alloy compositions is given by Soviet steel standards, GOST 4543-57 and 4632-51.

Constructional carbon steels of "ordinary" quality, which are specified according to their mechanical properties, are St0, St1, St2, St3, St4, St5, St6, and St7. The higher the figure, the greater the carbon content. Depending on the method of manufacture, the prefix B or M may be added to describe a Bessemer or open-hearth steel, respectively.

Constructional "quality" carbon steels have numbers between 05 and 70, which indicate the average carbon content in hundredths of one percent. These numbers may carry suffixes kp, sp, and ps, denoting rimmed, killed, and semikilled grades, respectively.

"Quality" carbon tool steels have designations U7, U8, etc., up to U12, the number indicating the average carbon content in tenths of one percent.

In alloy steel designations, the following symbols represent the alloying elements:

A	Nitrogen	Kh	Chromium	S	Silicon
B	Columbium	M	Molybdenum	T	Titanium
D	Copper	N	Nickel	Ts	Zirconium
F	Vanadium	P	Phosphorus	V	Tungsten
G	Manganese	R	Boron	Ya	Aluminum
K	Cobalt				

If the percentage of the element is not greater than about 1%, the letter for the element is not followed by a figure. If the amount of the element is greater than 1%, a figure representing the content is placed to the right of the letter, e.g., 4% Ni is represented by N4.

The average carbon content is shown to the left of the letters as hundredths of one percent. In the case of a very low carbon content (less than about 0.08%), the numeral 0 is placed before the letters. Occasionally, the carbon figure is omitted altogether.

The letter A appended to these designations denotes "high quality" (narrower composition limits and lower sulfur and phosphorus contents). This must not be confused with the symbol A representing nitrogen. Frequently, the letter L is used as a suffix to denote a cast steel.

Several groups of steel designations carry a prefixed letter which indicates a particular purpose or characteristic of the steel, e.g., A free-machining, E magnetic, Zh straight chromium stainless, R high-speed, Sh ball-bearing, E electrical, and Ya chrome-nickel stainless. The numbers following these letters are not normally indicative of the actual composition.

Many steels and alloys have designations consisting of the letters ÉI or EP followed by a number of up to three digits. Wherever possible, the actual compositions of these materials will be given.

PART I

DIFFUSION AND STRUCTURE OF METALS

DIFFUSION AND STRUCTURE OF METALS

S. Z. Bokshtein

Metal science deals essentially with crystalline substances and, therefore, whether the metal scientist studies the properties of metals or the primary processes occurring in metals he is concerned first of all with the structure.

Diffusion processes play a significant role in the formation of new phases, their development, and their disappearance, while the kinetics of diffusion is affected in turn by the structure of the metal and the phase transformations. The diffusion and structure of metals are closely related. The relationship between them has been studied in a number of investigations, in one of which the method of radioactive isotopes was used. The method of radioactive isotopes, and particularly the autoradiographic method, is a very effective method of investigating local displacements of atoms in a crystal lattice.

The study of the structure of real metals is largely the study of imperfections in the crystal lattice. Defects in the crystal structure lead to the creation of short channels, i.e., they facilitate diffusion. In this respect the surface of separation between the structural elements, particularly the grain boundaries, are of special importance.

Fisher [1] was the first to analyze the role of boundaries in the process of diffusion. He assumed that the grain boundary is a type of "fissure" perpendicular to the surface of separation and that the coefficient of diffusion along the grain boundaries is much larger than the coefficient of volume diffusion. Later, we proposed a model which is closer to a real crystal [2]. Unlike Fisher, we represented the crystal substance as some arrangement of spheroidal grains. The boundary between grains was described as a phase with specific equilibrium and kinetic characteristics. The diffusing substance diffuses into the two phases: the grain boundary and the body of the grain. Fisher calculated the width of the boundary on the basis of the theory concerning the region in which the forces of interatomic bonds act and found it to be equal to $5 \cdot 10^{-8}$ cm. We used the data obtained by Arkharov et al., and found a value two orders higher, i.e., $5 \cdot 10^{-6}$ cm.

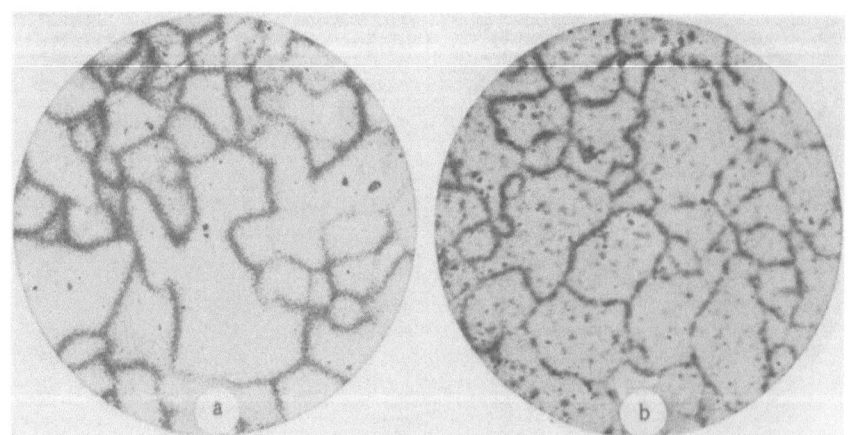

Fig. 1. Self-diffusion: a) iron at 700°C (autoradiogram). ×50. b) chromium at 1100°C (autoradiogram). ×30.

Activation Energy of Diffusion Along the Grain Boundaries and Within the Grain

Diffusing element	Dissolving metal	Activation energy Q, cal/g-atom	
		Q_g	Q_b
Fe	Fe	64,000	30,600
Ni	Ni	65,800	24,800
Cr	Cr	76,000	46,000
Mo	Mo	113,300	77,000
Sn	Ni	58,000	30,400
Sn	Ni + 0.1% B	59,000	35,000

EFFECT OF GRAIN BOUNDARIES

The imperfections in the structure of boundaries (dislocations, excess vacancies) result in excess energy which is assumed to be of the order of 2000-5000 cal/g-atom. Both the structure and the composition of the grain boundaries differ from those of the body of the grain. Investigation with the autoradiographic method [3] showed that grain boundaries are usually enriched in impurities and alloyed elements. The irregular distribution of impurities is detected by radioactive isotopes even when the amounts are extremely small—a thousandth or a millionth of one percent.

Systematic investigations of iron, nickel, chromium, molybdenum, and other alloys showed that the mobility of added elements along the grain boundaries is particularly clear in the case of self-diffusion. Self-diffusion for iron and chromium is shown in Fig. 1. This property of boundaries is also observed qualitatively in the case of heterodiffusion (Fig. 2). However, in the case of heterodiffusion the deformation of the crystal lattice resulting from the dissolution of foreign atoms also facilitates diffusion within the crystal.

The activation energy of self-diffusion along the boundary (Q_b) is about half that of self-diffusion within the grain (Q_g) (see the table).

In chromium the values of Q_b and Q_g and also Q_b/Q_g are higher than in nickel. This indicates that, from the thermodynamic and kinetic viewpoints, the difference between Q_b and Q_g is lower. In a similar way, the addition of 0.01% B to nickel or nickel alloys increases Q_b/Q_g.

Since $Q_s < Q_b < Q_g$ (representing the activation energy of diffusion on the surface, along the boundary, and within the grain, respectively), one can conclude that surface diffusion should occur at low temperatures, boundary diffusion should occur at medium temperatures, and only volume diffusion should occur at high temperatures. In reality, diffusion remains preferential along the grain boundaries up to very high temperatures: up to at least 1200°C in the case of iron and nickel, up to 1350°C in the case of chromium, and up to 1750°C in the case of the diffusion of tungsten in molybdenum. With the exception of the autoradiographic method, ordinary methods of investigation do not reveal this diffusion, since the contribution from the boundaries to the total diffusional flow at high temperatures, when the total diffusion rate is very high, is relatively small.

Systematic investigations by the autoradiographic method showed that in quite different metals and alloys the grain boundaries act as channels through which the atoms of different elements move preferentially.

However, titanium is an exception. Iron, chromium, and tin diffuse through the grain in titanium and (as a study of the microstructure showed) along certain planes (Fig. 3). One would assume that this type of diffusion is due to the structural peculiarities resulting from the $\beta \rightarrow \alpha$ transformation, which proceeds according to martensitic kinetics [4]. We confirmed this statement experimentally: after prolonged annealing (150 h at 750°C) leading to the creation of the equilibrium α-structure, diffusion proceeds preferentially along the grain boundaries (Fig. 4). However, we also found irregular diffusion within the grain, indicating structural heterogeneities.

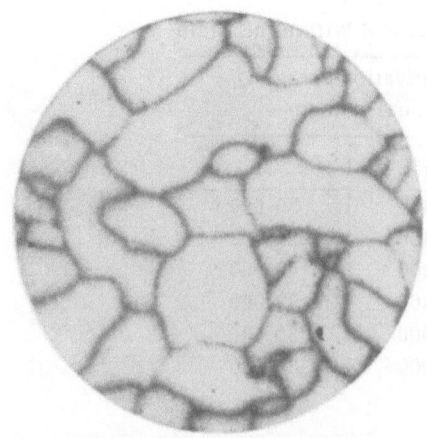

Fig. 2. Diffusion of tungsten in iron at 700°C (autoradiogram). ×50.

It should be noted that, contrary to widely held beliefs, diffusion proceeds along the grain boundaries even in interstitial solid solutions—the diffusion of carbon, for example, in α-iron (at temperatures of 500-950°C) and in γ-iron [5].

The effect of grain boundaries on the elementary act of diffusion is apparently related essentially to peculiarities in the structure of the boundary. Measurements of the diffusion coefficient of a tin alloy in nickel showed that the diffusion rate along the boundaries increases with the size of the grain [6]. Apparently, structural defects are more pronounced in large grains.

The irregular character of diffusion cannot be explained by the surface-active properties of the diffusion elements alone. This statement is confirmed by the great effect of the grain boundaries in the case of self-diffusion (see Fig. 1) and also by autoradiographic studies [6], which showed that the diffusion of chromium in iron as well as the diffusion of iron in chromium or mutual diffusion in the iron-nickel system is predominantly boundary diffusion.

EFFECT OF PHASE TRANSFORMATIONS

The short channels and the increase in diffusional mobility may occur at the moment of phase transformations.

We have shown earlier [7] that the polymorphic $\alpha \rightleftharpoons \gamma$ transformation in iron and the eutectoid austenite-pearlite transformation in steel (0.08% C) affect the self-diffusion of iron. The samples were subjected to diffusional annealing at a constant temperature in one case and to alternate heating and cooling in the other case. In the latter case the self-diffusion of iron is accompanied by polymorphic and eutectoid transformations.

The results show that polymorphic transformations have no great effect on the elementary act of diffusion and that two processes can be considered to be independent. Apparently, the atoms do not move any significant distance when one type of stacking of iron atoms is replaced by another. On the other hand, under the condition of a complex eutectoid transformation (austenite $\rightleftharpoons$ pearlite) the self-diffusion coefficient increases by a factor of 10.

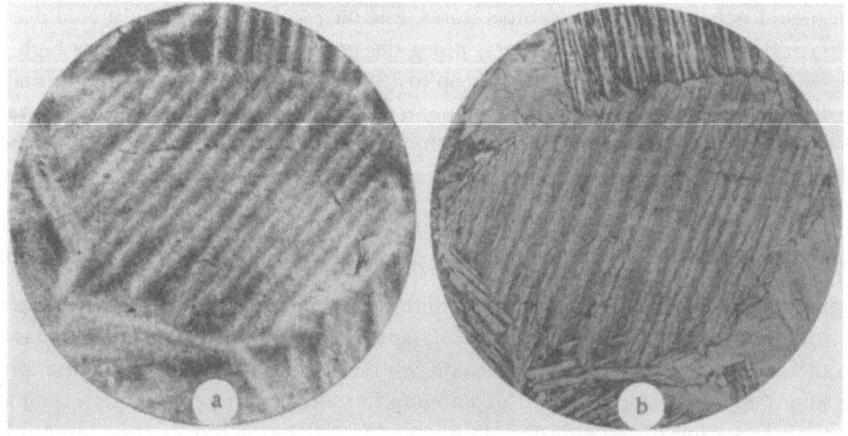

Fig. 3. Diffusion of iron in titanium at 800°C. ×30. a) Autoradiogram; b) microstructure.

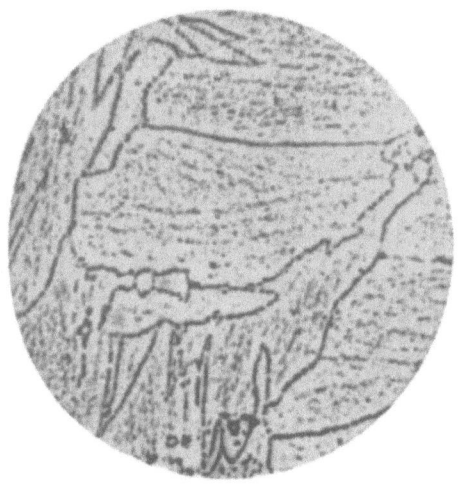

The formation of a new phase and the resulting reconstruction of the lattice during the eutectoid transformation may lead to an increase in mobility for the following reasons: the crystal lattice can loosen at the surface of separation between the mother phase and the new phase and the phase transformations may be accompanied by phase cold hardening, leading to an increased diffusion rate.

Investigation of titanium alloys showed that recrystallization processes may lead to an increase in diffusional mobility.

However, the diffusion rate is lower after the phase transformation. Thus, the investigation of the Kh20N80T3 (ÉI437) nickel alloy showed that after aging is terminated the diffusion rate of tin along the grain boundaries decreases by a factor of five with respect to the diffusion rate in the quenched alloy.

Fig. 4. Diffusion of nickel in titanium at 750°C (autoradiogram). ×50.

EFFECT OF THE TYPE OF CRYSTAL LATTICE

It is well known that the self-diffusion rate of iron in the γ-state is considerably higher than in the α-state. For the same value of activation energy of self-diffusion (about 70,000 cal/g-atom) the coefficient of self-diffusion at the temperature of the transformation of iron is two to three orders higher than in γ-iron [8].

In this respect titanium is particularly interesting. In spite of the high melting point, the diffusional mobility in titanium is relatively high. We compared the diffusion of tin in α- and β-titanium at temperatures of 700-1100°C to prove this point [9].

We could not study the self-diffusion of titanium by the radioactive isotopes method because there are no radioactive isotopes of titanium, and tin is neutral with respect to the polymorphic transformation in titanium. The diffusion (D) decreases sharply during the transformation of α-titanium into β-titanium. Extrapolation to the transformation temperature showed that the diffusion coefficient in α-titanium is about 100

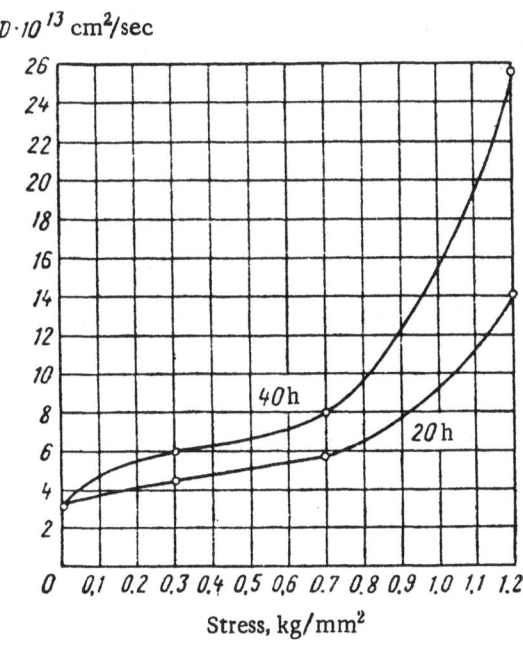

Fig. 5. Effect of stress on the self-diffusion coefficient of iron at 750°C.

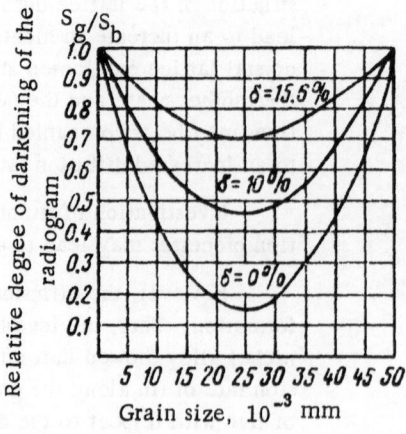

Fig. 6. Distribution of diffusional flow in deformed and
undeformed samples (diffusion of tin in nickel at 800°C).

times higher than in β-titanium, while the activation energy is lower by a factor of 2.5. In technically pure
α- and β-titanium the temperature dependence is of the form:

$$D_\alpha = 8.9 \cdot 10^{-4} \cdot e^{\frac{38000}{RT}},$$

$$D_\beta = 10^4 \cdot e^{\frac{86500}{RT}}.$$

Thus, the empirical relationship $Q = 40 T_s$ (where Q is the activation energy of diffusion and T_s the melting point) holds in the case of β-titanium.

It is assumed that the diffusional mobility in body-centered cubic iron is higher than in face-centered cubic iron because the atoms in α-iron are much less densely stacked. This explanation does not hold in the case of titanium, since the atoms are more densely stacked in α-titanium than in β-titanium. We assume that diffusion is facilitated in the low-temperature modification of titanium in the α'-state because of the formation of inner separation surfaces resulting from the $Ti_\beta \rightarrow Ti_\alpha$ transformation. The diffusion of tin, chromium, and iron in titanium after polymorphic transformation (see Fig. 3) and the low value of D_0 for α-titanium seem to confirm this assumption.

EFFECT OF STRESS AND DEFORMATION

The effect of external forces, particularly stress and deformation, on diffusion is of practical and theoretical importance. The study of this process by the absorption method and by the autoradiographic method showed that tensile stresses and deformation considerably increase the rates of diffusion and self-diffusion[10].

The coefficient of self-diffusion of iron and the coefficient of diffusion of tin in nickel alloys increase by a factor of two to three under the effect of low elastic-plastic deformation and increase tens of times under the effect of plastic deformation. Thus, at 750°C ($\sigma = 1.2$ kg/mm^2, $\delta = 18\%$) the self-diffusion coefficient of iron increases from 3.2 to $26 \cdot 10^{-13}$ cm^2/sec (Fig. 5).

The relative change in the diffusion coefficient as the result of stress decreases with increasing temperature. The increase in the diffusion rate is due to the decrease in the activation energy of this process. For example, in the case of self-diffusion of iron it decreases from 69,900 cal/g-atom in the absence of 46,300 cal/g-atom under a stress $\sigma = 0.3$ kg/mm^2 and $\delta = 10\%$.

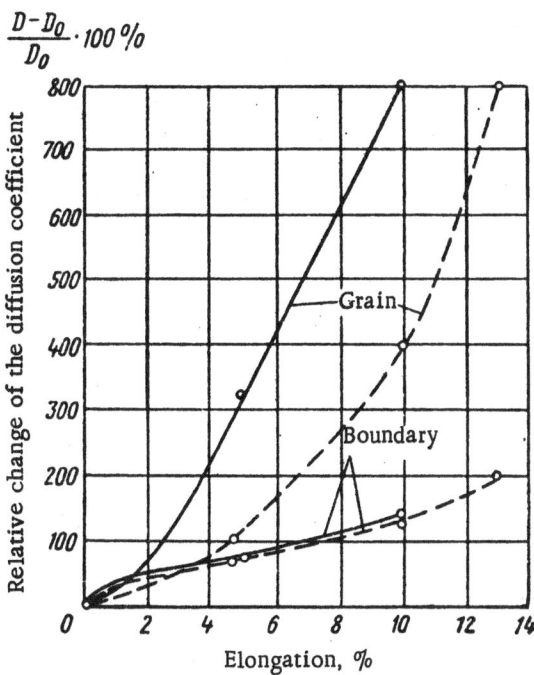

$$\frac{D-D_0}{D_0} \cdot 100 \%$$

Fig. 7. Effect of preliminary plastic deformation on the diffusion of tin within the grain and along the grain boundaries at 800°C. —) Cold deformation; ---) hot deformation (700°C).

The character of the diffusional flow changes under the effect of plastic deformation, as is shown by the autoradiographs. From predominantly boundary diffusion it becomes volume diffusion with increasing deformation (Fig. 6). Thus, the change in the kinetics of the diffusion process is explained to a great extent by the irreversible structural changes induced by plastic deformation. The investigations of the effect of preliminary plastic deformation on the diffusion rate lead to the same conclusion. Plastic deformation at room temperature and at high temperature (700°C) induce an increase in the diffusion coefficient of tin in nickel along the boundaries and, particularly, within the grain (Fig. 7). It was shown that the changes induced by preliminary plastic deformation are very stable and may remain even after the metal is recrystallized.

From a number of investigations with radioactive isotopes we determined some characteristic relationships between the primary process of diffusion, phase transformation, and the structure of the metal. These relationships must be taken into account in predicting the behavior of alloys at high temperatures and under high stress.

Many problems concerning the relationships between the elementary process of the displacement of atoms in the crystal lattice and the structure of the lattice require further study.

LITERATURE CITED

1. J. C. Fisher, J. Appl. Phys. 22 (1): 74, 1951.
2. B. S. Bokshtein, I. A. Magidson, and I. L. Svetlov, Fiz. Metal. i Metalloved. 6:1036, 1958.
3. S. Z. Bokshtein, S. T. Kishkin, and L. M. Moroz, Investigation of the Structure of Metals by the Radioactive Isotopes Method, Oborongiz, 1959.
4. S. Z. Bokshtein, S. T. Kishkin, and V. B. Osvenskii, Metalloved. i Term. Obrabotka Metal. No. 6, 1960.
5. S. Z. Bokshtein, M. A. Gubareva, I. E. Kantorovich, and L. M. Moroz, Metalloved. i Term. Obrabotka Metal. No. 1, 1961.

6. S. Z. Bokshtein, S. T. Kishkin, and L. M. Moroz, Investigation of the Structure of Metals by the Radio-active Isotope Method, Oborongiz, 1959.

7. S. Z. Bokshtein, A. A. Zhukhovitskii, S. T. Kishkin, and É. R. Mal'tsev, Nauchn. Dokl. Vysshei Shkoly, No. 4, 1958.

8. C. Birchenall and R. Mehl, J. Metals 188 (1): 144, 1950.

9. S. Z. Bokshtein, S. T. Kishkin, and V. B. Osvenskii, Metalloved. i Term. Obrabotka Metal. No. 6, 1960.

10. S. Z. Bokshtein, T. I. Gudkova, A. A. Zhukhovitskii, and S. T. Kishkin, Some Problems of the Strength of Solids, Izd. Akad. Nauk SSSR, 1959.

DIFFUSION AND SELF-DIFFUSION IN NICKEL ALLOYS

M. A. Gubareva and L. M. Moroz

Diffusion and self-diffusion are the decisive processes determining the stability of the structure and the properties of alloys at high temperatures.

Heat-resistance alloys usually crack along the grain boundaries, and therefore the state of the boundaries and the diffusional mobility along the grain boundaries are of great importance.

Nickel alloys are the principal heat-resistant materials used for the most critical machine parts, and therefore it is important to study self-diffusion and diffusion in these alloys as a function of their composition and structure.

For this study we used the previously developed autoradiographic method on samples cut at an angle to the surface coated with radioactive material. With this method it is possible to determine the parameters of volume and boundary diffusion separately.

A number of investigations have been published [1-6] on the diffusion of nickel in nickel and in nickel alloys. An attempt was made in [6] to relate the diffusional characteristics to the heat-resistance characteristics of these alloys. This study was concerned primarily with the parameters of volume diffusion in nickel. In [7] the self-diffusion of nickel along the grain boundaries was investigated as a function of the orientation of neighboring grains.

It was shown there that the depth of penetration of radioactive nickel along the boundaries depends on the orientation of grains and that the maximum depth of penetration corresponds to the maximum degree of disorientation (Θ).

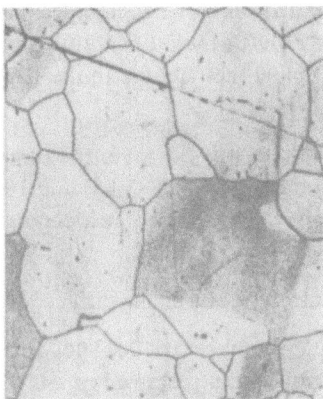

Fig. 1. Self-diffusion of nickel at 700°C. Autoradiogram. ×50.

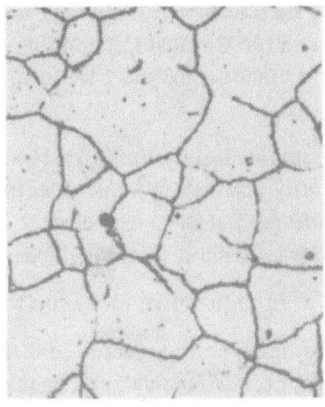

Fig. 2. Self-diffusion of nickel at 800°C. Autoradiogram. ×50.

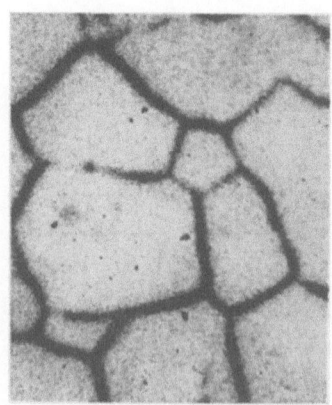

Fig. 3. Self-diffusion of nickel
at 1000°C. Autoradiogram. ×50.

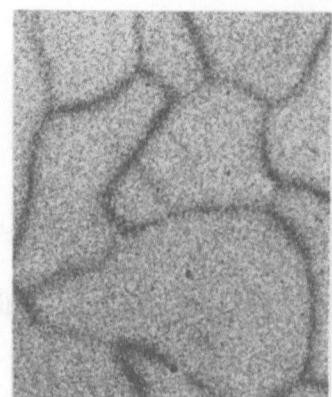

Fig. 4. Self-diffusion of nickel
at 1200°C. Autoradiogram. ×50.

The ratio between the diffusion coefficient along the grain boundaries D_b and the diffusion coefficient within the grain D_g varies from 10^3 to 10^7 when Θ varies from 5 to 80°C. This ratio increases with decreasing temperature.

The activation energy of self-diffusion along the grain boundaries is 26,000 cal/g–atom, i.e., lower by a factor of 2.5 than the activation energy of diffusion within the grain, which is 65,800 cal/g–atom. The activation energy of self-diffusion remains constant at disorientation angles of 20-70°; at lower angles and at $\Theta = 0$ the activation energy is equal to 65,900 cal/g–atom, i.e., is practically the same as the activation energy of volume diffusion.

It should be noted that there are relatively few data on the diffusion and self-diffusion of nickel and that some of these data are contradictory. There are no data on self-diffusion of nickel along the grain boundaries in technically pure nickel or nickel alloys.

We investigated the diffusion of nickel by the autoradiographic method and in some cases we determined the parameters of the self-diffusion of nickel in nickel alloys.

MATERIALS AND METHOD

We investigated technically pure nickel containing 0.01% B and also nickel alloyed with tungsten (0.6%). We also investigated the diffusion of nickel in the heat-resistant ÉI437A nickel alloy.

These materials were subjected to high-temperature annealing as follows. Technically pure nickel was heated 1 h at 1100°C; nickel containing 0.01% B was heated 12 h at 1100°C; nickel containing 0.6% W was heated 9 h at 1200°C; and the ÉI437A alloy was heated 2 h at 1200°C and then cooled in the furnace to 900°C and then in air.

The samples were prisms 10 × 10 × 20 mm. To eliminate the cold-hardened layer the samples were electrolytically polished in Jaquet electrolyte: 50 ml of hydrochloric acid (d = 1.58-1.61) and 1000 ml of glacial acetic acid at 30°C, the density of current being 20-30 A/dm^2. The distance between the anode and the cathode was 20-25 mm; the cathode was aluminum.

A layer of radioactive nickel up to 1 μ thick was deposited electrolytically on the surfaces of the samples.

The samples were subjected to diffusional annealing in a vacuum furnace at 700, 800, 1000, and 1200°C. The temperature was recorded and checked with the EPD-17 apparatus, with a precision of ±2°.

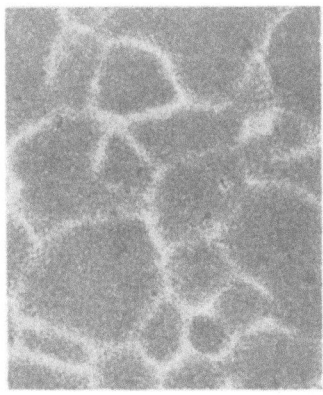

Fig. 5. Autoradiogram of the surface of the sample after diffusional annealing at 800°C. ×50.

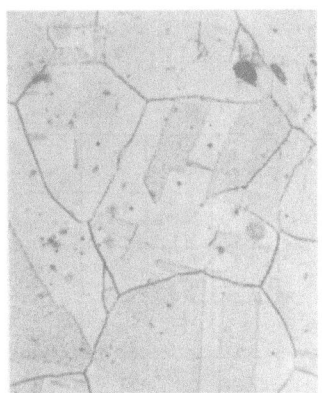

Fig. 6. Microstructure of technically pure nickel. ×50.

RESULTS

Qualitative Description of the Process

Figures 1-4 show autoradiograms of nickel samples after diffusional annealing at 700, 800, 1000, and 1200°C, respectively. One can see clearly that the self-diffusion of nickel occurs essentially along the grain boundaries in all cases, even at 1200°C. At the same time, one can note the effect of temperature on the character of this process; at 1200°C there is also considerable diffusion through the grains.

These autoradiograms show that the diffusional zone along the grain boundaries changes from very thin at 700°C (Fig. 1) to rather wide and washed out at 1200°C (Fig. 4).

Figure 5 shows an autoradiogram of the surface of the sample after diffusional annealing at 800°C; the light grain boundaries surrounding the dark grains are clearly visible. Apparently this autoradiogram is characteristic of the diffusional flow at the initial moment when the diffusing substance penetrates the metal along the grain boundaries, and as the result the boundaries at the surface are greatly impoverished in the radioactive substance.

Let us note that the autoradiogram obtained with Ni^{63} is particularly clear. This is due to the fact that nickel is essentially a low-energy β-radiator. The autoradiograms obtained, for instance, after diffusional annealing at 700°C are exactly the same as the micrograph (Fig. 6).

Self-diffusion of nickel occurs essentially along the grain boundaries in the nickel alloy containing 0.01% B, in nickel containing 0.6% W, and in pure nickel. In the nickel alloy containing boron the process occurs essentially along the grain boundaries up to 1200°C, although diffusion through the grains also occurs.

In the nickel alloy containing tungsten the clearly expressed boundary diffusion at 800°C is accompanied in some cases by diffusion through the grains, which does not occur in other alloys at this low temperature.

In ÉI437A and ÉI437B nickel alloys as well as in pure nickel and in binary nickel alloys self-diffusion at 800-1200°C is essentially along the grain boundaries.

Quantitative Evaluation of the Process

The parameters of the self-diffusion of nickel in pure nickel and in Kh20N80T3 (ÉI437) were determined by the autoradiographic method on samples cut at an angle to the surface coated with radioactive material.

The coefficients of diffusion along the boundaries and through the grains were calculated on the assumption that these processes are independent. We did not take into account the migration of the atoms of the diffusing substance from the boundaries into the grains.

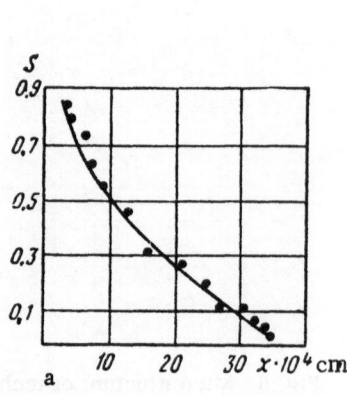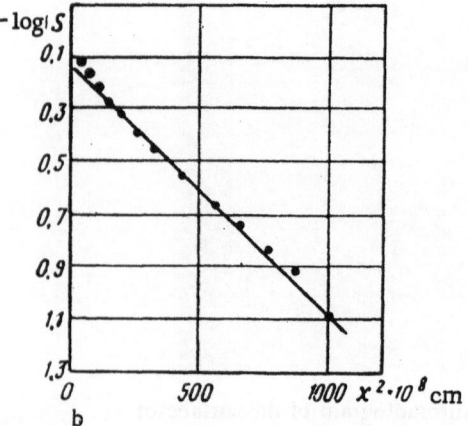

Fig. 7. Variation of the degree of darkening of the autoradiogram with the depth of self-diffusion of nickel along the grain boundaries at 1000°C. a) $S = f(x)$; b) $S = \psi(x^2)$.

In Figs. 7a and 7b we show the dependence of the degree of darkening on the depth of self-diffusion at 1000°C and also the variation of the logarithm of the degree of darkening on the square of the depth at which the diffusion coefficients were calculated.

The coefficients of self-diffusion of nickel along the boundaries in pure nickel and in the Kh20N80T3 alloy at 800, 1000, and 1100°C are given in Table 1.

No comparison of the results in Table 1 is possible because, as far as we know, there are no other published data on the self-diffusion of nickel along the grain boundaries in nickel or the Kh20N80T3 alloy. Comparison of these results with the self-diffusion of iron along the grain boundaries at 800°C (which we determined previously as $1 \cdot 10^{-11}$) shows that the coefficient of self-diffusion of nickel along the grain boundaries at 800°C is one-fifth of the coefficient of self-diffusion of iron.

The activation energy of the self-diffusion of nickel along the grain boundaries was found to be 24,800 cal/g-atom.

According to data published abroad [7], $Q_b = 26,000$ cal/g-atom, which is in good agreement with our results.

Comparison of the coefficients of self-diffusion of nickel in pure nickel and in a nickel alloy shows that the diffusional mobility along the grain boundaries in the nickel alloy is much lower than that in pure nickel. For example, the coefficient of self-diffusion of nickel in the Kh20N80T3 alloy (ÉI437) at 800°C is lower by a factor of 3.5 than in pure nickel—$0.6 \cdot 10^{-12}$ as compared with $2.0 \cdot 10^{-12}$ cm²/sec. At 1200°C the coefficient of boundary diffusion in the Kh20N80T3 alloy is about half that in pure nickel. The activation energy of self-diffusion of nickel along the grain boundaries in this alloy turns out to be 30,600 cal/g-atom, which is considerably higher than in pure nickel.

Thus, by using autoradiograms we have shown that the self-diffusion of nickel in pure nickel is irregular and the rate of self-diffusion is greater along the grain boundaries. This conclusion agrees with results obtained for other metals (iron and chromium, for example).

The preferential self-diffusion of nickel along the grain boundaries occurs not only in pure metals but also in nickel alloys: Ni + B, Ni + W, and also in the complex alloy Kh20N80T3.

However, in alloys the effect of boundaries is not so sharp as in pure nickel, apparently because dissolution of atoms during the formation of solid solutions leads to the formation of defects within the grain and, consequently, to the creation of areas of high mobility.

TABLE 1

Temperature of diffusional annealing, °C	Self-diffusion coefficient of nickel along the grain boundaries, $D_b \cdot 10^{12}$, cm²/sec	
	Nickel	Kh20N80T3 alloy (ÉI437)
800	2.0	0.6
1000	11.0	4.8
1200	56.0	28.0

TABLE 2

Temperature of diffusional annealing, °C	Diffusion coefficient, $D \cdot 10^{13}$, cm²/sec	
	Cast alloy	Forged alloy
800	0.3	0.8
850	2.0	5.8
950	23.0	33.0

TABLE 3

Temperature of diffusional annealing, °C	Diffusion coefficient, $D \cdot 10^{13}$, cm²/sec			
	Cast alloy		Forged alloy	
	D_g	D_b	D_g	D_b
800	0.3	—	0.7	3.0
850	1.7	7.0	2.6	6.2
950	22.0	45.0	20.0	34.0

TABLE 4

Diffusing element	Solvent	Diffusion coefficient, $D \cdot 10^{13}$, cm²/sec	
		D_g	D_b
Nickel	Nickel	—	20.0
Nickel	Kh20N80T3 alloy	—	5.8
Tin	Nickel	13.0	210.0
Tin	Ni + 0.01% B	6.3	63.0
Tin	Kh20N80T3 alloy	4.0	60.0
Tin	Kh20N80T3 alloy + 0.01% B	3.2	20.0
Tin	Forged ZhS3 alloy	0.7	3.0
Tin	Cast ZhS3 alloy	0.3	—

TABLE 5

Diffusing element	Solvent	Q, cal/g–atom		Q_b
		Q_g	Q_b	$\dfrac{Q_b}{Q_g}$
Nickel	Nickel	65900 [7]	24800	0.38
Nickel	Kh20N80T3 alloy	–	30600	–
Tin	Nickel	58000	30400	0.52
Tin	Ni + 0.01% B	59000	35000	0.60
Tin	Kh20N80T3 alloy	65000	40500	0.62
Tin	Kh20N80T3 alloy + 0.01% B	65400	48600	0.74
Tin	Forged ZhS3 alloy	65000	49000	0.75
Tin	Cast ZhS3 alloy	70000	52000	0.75

This is particularly clear in the nickel alloy containing tungsten. Tungsten is located essentially within the grains and, because of the great difference of the atomic radii of tungsten and nickel, the crystal lattice is greatly deformed.

The boundaries still affect the self-diffusion of nickel even at very high temperatures (up to 1200°C for pure Ni, Ni + B, and the Kh20N80T3 alloy). We did not investigate temperatures above 1200°C. However, the role of volume diffusion increases with the temperature, as can be seen from our data, and the width of the boundary diffusion zone increases. Apparently, the boundaries become looser and the width of the transition zone from a more perfect structure (of the grain) to a less perfect structure (of the boundary) increases with increasing temperature.

The rate of self-diffusion of nickel in pure nickel is considerably lower than the rate of self-diffusion of iron in pure iron under the same conditions. Similarly, the self-diffusion coefficient along the grain boundaries in the Kh20N80T3 alloy is lower than in pure nickel, and the lower the temperature the lower the coefficient. Thus, at 800°C D_b for Kh20N80T3 is one-quarter of that for pure nickel, while at 1200°C it is half that for pure nickel.

EFFECT OF THE STRUCTURE OF STABLE ALLOYS ON THE DIFFUSION RATE

We studied the effect of the structure on the diffusional mobility in the ZhS3 nickel alloy. In particular, we investigated the mobility of tin in cast and forged alloys.

The cast and forged alloys were heated 7 h at 1150°C and cooled in air. The samples were then subjected to diffusional annealing at 800, 850, and 950°C in an argon atmosphere. The temperatures of diffusional annealing of 800 and 850°C are the operating temperatures of the alloy; at 950°C the alloy loses its strength.

We determined quantitatively the diffusion rate of tin in cast and forged ZhS3 alloys. We determined the average value of the diffusion coefficients (by the average degree of darkening of the autoradiograms) and also the diffusion rates along the boundaries and through the grains.

The average values of the diffusion coefficient of tin in cast and forged ZhS3 alloys are shown in Table 2. The coefficients of boundary and volume diffusion are given in Table 3.

Plastic deformation changes the structure of the alloy, shortens the diffusion channels, and increases the total length of the boundaries, i.e., the areas of high mobility. Consequently, the concentration of the components in the forged alloy becomes uniform much more rapidly, since there is a great surface of separation in this alloy than in the cast alloy. The average diffusion coefficients confirm this statement. At all temperatures investigated, the average diffusion coefficients are greater for the forged than for the cast alloy (see Table 2).

The comparison of the diffusion coefficients through the grain and along the grain boundaries for cast and forged alloys (see Table 3) shows that at 800 and 850°C the diffusion rate within the grain in the forged alloy is greater than that for the cast alloy, while at 950°C the diffusion coefficients are about the same in both types of alloys. The fact that the rates of volume diffusion are the same at 950°C can be explained by the high rate of diffusion at this temperature, which is not affected by the structure of the alloy.

The diffusion rates along the grain boundaries in cast and forged alloys are approximately the same, although somewhat higher in the cast alloy. This result agrees with the results of our investigation of the effect of the grain size of nickel on the rate of boundary diffusion, which showed that the rate of diffusion along the boundaries increases by a factor of about seven when the grain size of nickel increases by a factor of about five.

In this case the size of the grains in the cast ZhS3 alloy is about five times greater than the size of the grains in the forged alloy.

Apparently the structure of the boundary becomes less ordered with increasing grain size. It is also possible that the width of the boundary increases or that the rate of transfer of substances from the boundary into the grain decreases. As the result, the diffusion rate along the boundaries of a large-grained material is higher than in a small-grained material.

Table 4 contains the data on the self-diffusion of nickel in pure nickel and in the Kh20N80T3 alloy and also the data on the diffusion of tin in pure nickel and in the Kh20N80T3 and ZhS3 alloys at 800°C.

The table shows that the rate of volume diffusion of tin in the Kh20N80T3 alloy is one-third the diffusion rate in pure nickel; the rate of boundary diffusion of tin in the Kh20N80T3 alloy is lower by a factor of 3.5 than the diffusion rate in pure nickel; the rates of diffusion of tin in the ZhS3 alloy are 40 and 70 times lower, respectively. The depth of diffusion in the Kh20N80T3 alloy is also less than in pure nickel. In the ZhS3 alloy heated 120 h at 800°C the depth of the diffusion of tin is 5-7 μ within the grain and 10-12 μ along the grain boundaries, while the depth of diffusion in pure nickel at the same heating time and temperature is 120-140 μ.

It should be noted that the activation energy (Q) of the diffusion process increases with the degree of alloying (Table 5). It can be seen in Table 5 that the activation energies of the diffusion of tin through the grain and along the grain boundaries of the cast ZhS3 alloy are greater than in pure nickel or in the Kh20N80T3 alloy. Apparently this is also one of the causes of the higher heat resistance of the ZhS3 alloy as compared to the Kh20N80T3 alloy.

The ratio of Q_b/Q_g is also larger for the ZhS3 alloy than for the Kh20N80T3 alloy or pure nickel (0.75, 0.62, and 0.52, respectively). This indicates that in the ZhS3 alloy there is a greater kinetic and thermodynamic similarity between the grain boundary and the grain than in the Kh20N80T3 alloy and greater still than in pure nickel.

Thus, we have shown for the first time that in nickel alloys, which are the principal heat-resistant materials, self-diffusion is a heterogeneous process. We have shown the role of the boundaries, temperature, composition, and structural state in the processes of self-diffusion and heterodiffusion.

LITERATURE CITED

1. R. Hoffman, F. Pikus, and N. Word, Trans. AIME 206:483, 1956.
2. I. Reinolds, B. Averbach, and M. Cohen, Acta Met. 5:29, 1957.
3. J. R. McEvan, J. U. McEvan, and L. Jaffe, J. Can. Chem. 37(10):1623, 1959.
4. H. Burgers and R. Smoluchowski, J. Appl. Phys. 26:491, 1955.
5. J. R. McEvan, J. U. McEvan, and L. Jaffe, J. Can. Chem. 37(10):1629, 1959.
6. S. D. Gertsriken, et al., Collection of Reports on the Theory of Heat Resistance, Izd. Akad. Nauk SSSR, 1961, 50-56.
7. W. R. Upthegrove and M. J. Sinnott, Trans. ASM 50:1031-1046, 1958.

VOLUME AND BOUNDARY DIFFUSION
OF TUNGSTEN IN MOLYBDENUM

S. Z. Bokshtein, M. B. Bronfin, and S. T. Kishkin

Tungsten and molybdenum have a number of similar physical and electrochemical properties (type of lattice, atomic radius, valence, etc.) and therefore one may assume that their diffusion coefficients in the same solvent are similar.

The study of the diffusion of tungsten in molybdenum is interesting not only in view of checking this assumption but also because direct study of intercrystalline self-diffusion of molybdenum is difficult—the half-life of the Mo^{99} isotope is very short and the β-radiation is hard.

W^{185} is for practical purposes a pure β-radiator, the maximum energy of electrons being 0.428 MeV and the half-life 74.5 days. The use of this isotope makes it possible to determine separately the boundary and volume diffusion coefficients in molybdenum by the autoradiographic method.

It should also be noted that information concerning the diffusion of tungsten in molybdenum is of interest for practical applications of alloys of these metals.

VOLUME DIFFUSION OF TUNGSTEN IN MOLYBDENUM

We investigated the volume diffusion of tungsten in molybdenum by two methods: by shaving off successive layers in the diffusion zone (by the shift of the activity curve [1]) and by the autoradiographic method. The molybdenum used for the measurements of the diffusion coefficient of tungsten by the shift of the activity curve had a large-grained structure, the average diameter of the grains being 0.5-1 mm. Before depositing the radioactive substance on the samples, one end of the cylindrical samples (14 mm in diameter) was electrolytically polished in a solution consisting of seven volumes of ethyl alcohol and one volume of concentrated sulfuric acid.

W^{185} was electrolytically deposited on the electropolished surfaces of the samples from a bath composed of 150 g/liter of ammonium sulfate, 60 g/liter of an aqueous solution of ammonia, and the radioactive anhydride $W^{185}O_3$ with an activity of ∼2-3 mCi.*

The anode was platinum, its area was several times larger than the activated area of the sample. The radioactive isotope was deposited at 20-40°C, the density of current being 1 A/cm^2, for a time sufficient for a total radioactivity in the deposited layer of 10,000-15,000 decompositions per minute. The activated samples were subjected to diffusional annealing in a high-temperature TVV-2 electric vacuum furnace with a tungsten heating element. The annealing temperature was measured with a VR 5/20 tungsten-rhenium thermocouple. Microlayers from the surface covered with the radioactive material were removed automatically with a specially constructed apparatus which combined mechanical rubbing and anodic dissolution [2]. The radioactivity was counted during removal of the layers with a β-counter, using the radiometric VSP apparatus.

---

* The composition prescribed by the Central Scientific Institute of Ferrous Metallurgy.

The diffusion coefficients of tungsten in molybdenum determined by the shift of the activity curve under different conditions of diffusional annealing are given in the following table:

Conditions of diffusional annealing	1830°−47.5 h	1880°−109.5 h	1950°−59 h	2100°−24 h
Diffusion coefficient, D, cm²/sec	$5.8 \cdot 10^{-12}$	$1.1 \cdot 10^{-11}$	$3.1 \cdot 10^{-11}$	$1.25 \cdot 10^{-10}$

The samples in which the diffusion of tungsten was investigated by the autoradiographic method were flat plates $10 \times 15 \times 3$ mm and had a small-grained structure. A layer of W^{185} was deposited on one surface of the sample in the manner described above. After annealing, samples were cut at an angle of 10-30' to this surface. The cut surface of the sample was placed on an NIKFI photographic plate of the MR type to obtain autoradiograms. The exposures varied from several days to one month, depending on the level of radioactivity in the sample. The autoradiograms obtained indicate the distribution of the radioactive tungsten in the surface layer (where diffusion occurs) and, as was shown before, the degree of darkening of the autoradiogram is directly proportional to the concentration of W^{185} on the surface of the sample.

Figure 1 shows one of the autoradiograms. This autoradiogram shows the diffusional distribution of W^{185} within the volume and along the grain boundaries of the sample annealed 108 h at 1750°C. In this sample the diffusional layer was cut at an angle of 15' to the activated surface. Preferential diffusion along the grain boundaries is clearly visible in the figure. To calculate the diffusion coefficient D after a given annealing time τ, we used the relationship

$$S = \text{const} \exp(-x^2/4D\tau), \qquad (1)$$

where S is the degree of darkening of the autoradiogram, which is proportional to the concentration of the radioactive isotope, and corresponds to the thickness x of the diffusion layer. The degree of darkening of the volume of the grains was recorded on the MF-4 microphotometer at several points on lines of equal depth perpendicular to the direction of the cut surface. Thus, each degree of blackening S substituted into Eq. (1) was an arithmetic average of 8-15 measurements for a given thickness x.

The variation of the degree of darkening of the volume of the grain on the autoradiogram with the thickness of the diffusion layer follows relationship (1) satisfactorily, as Fig. 2 shows. The respective coefficients of diffusion can be determined from the tangents of the slopes of the lines similar to that shown in Fig. 2.

The coefficients of volume diffusion of tungsten in molybdenum were determined by the autoradiographic method in samples annealed under the following conditions:

Conditions of diffusional annealing	1700°−112 h	1750°−108 h	1850°−103 h	1900°−99 h
Diffusion coefficient, D, cm²/sec	$9.9 \cdot 10^{-13}$	$2.0 \cdot 10^{-12}$	$8.9 \cdot 10^{-12}$	$1.2 \cdot 10^{-11}$

The temperature dependence of the diffusion coefficients of tungsten in molybdenum is shown in Fig. 3. The experimental points determined by the shift of the activity curve and also autoradiographically follow rather well the same straight line drawn in log D and 1/T coordinates, and this can be considered as additional proof of the validity of the method of the shift in the activity curve.

Applying the method of least squares to the coefficients obtained, we find the following equation for the diffusion of tungsten in molybdenum:

$$D = 3.18 \exp\left[-(112900 \pm 1000)/RT\right] \text{ cm}^2/\text{sec.} \qquad (2)$$

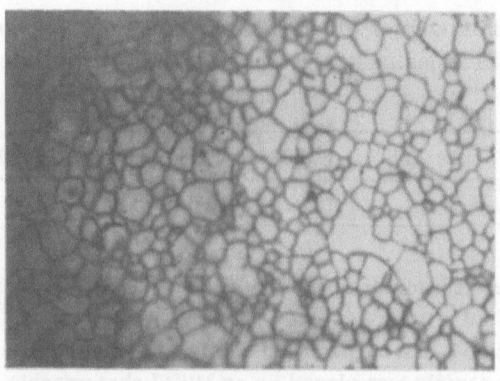

Fig. 1. Autoradiogram of the surface of molybdenum cut
at an angle of 15' to the surface coated with W^{185} and annealed 108 h at 1750°C. ×13.

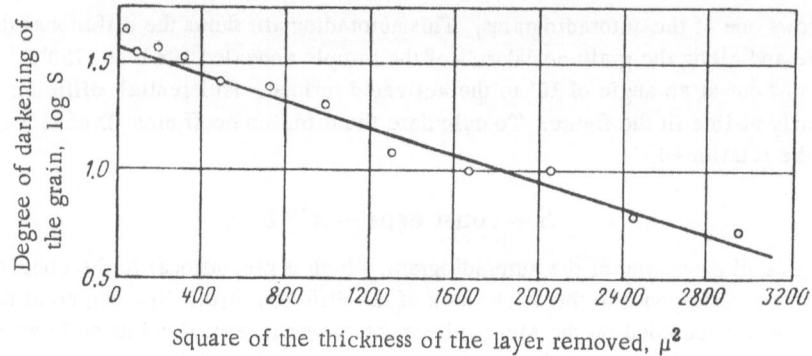

Fig. 2. Variation of the degree of darkening (S) of the grain on the autoradiogram with the thickness of the layer removed (diffusion of W^{185} in molybdenum
at 1900°C for 99 h).

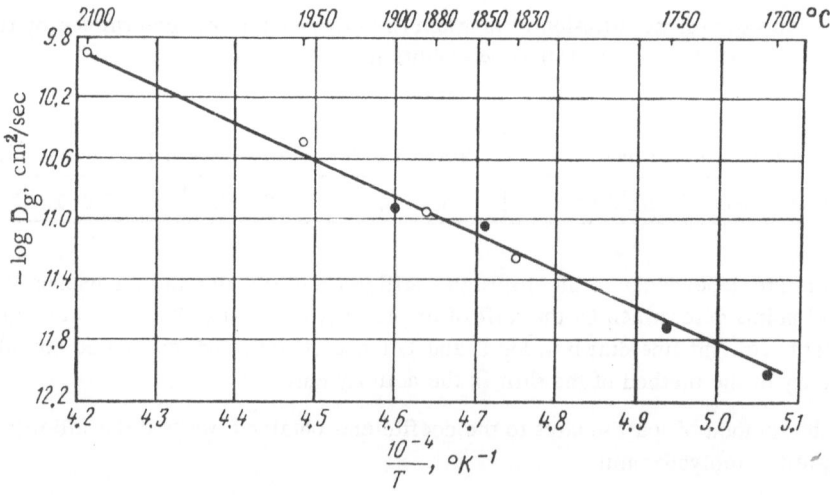

Fig. 3. Temperature dependence of the coefficient of volume diffusion of
tungsten in molybdenum. Method of measuring: ●) autoradiographic; ○) shift.

Impurity	Solvent	Atomic diameter of impurity, kX (coordination number 12)	Activation energy, kcal/g−atom	Source of data
Nickel	Nickel	2.49	66.8	13
Cobalt		2.50	68.3	14
Cobalt	Cobalt	2.50	67.7	15
Nickel		2.49	72.1	16
Silver	Silver	2.88	45.7	17
Gold		2.88	45.5	18
Gold	Gold	2.88	45.3	19
Silver	Lead	2.88	15.2	20
Gold		2.88	14.0	20
Niobium	Niobium	2.94	105	21
Tantalum	Tantalum	2.94	110	22
Antimony	Germanium	3.23	55.6	23
Arsenic		2.92	55.6	23
Molybdenum	Molybdenum	2.80	113	24
Tungsten		2.82	112.9	24

Equation (2) differs considerably from the temperature dependence of the diffusion coefficient of tungsten in molybdenum derived in [3]:

$$D = 5 \cdot 10^{-4} \exp(-78000/RT) \text{ cm}^2/\text{sec.} \qquad (3)$$

The lower value of the activation energy of the diffusion of tungsten in molybdenum is probably due to the fact that in [3] the authors did not succeed in excluding the effect of boundary diffusion.

Our value for the activation energy of the diffusion of tungsten in molybdenum is practically identical with the value of the activation energy of the self-diffusion of molybdenum (113,000 cal/g-atom) [4]. Our result apparently confirms the general assumption that in the case of the vacancy diffusion mechanism of atoms in a given solvent the activation energy of diffusion is determined basically by the valence and by the shape factor of the elements considered. When the valence, the atomic diameter, and other physical properties of the substances diffusing in a given solvent are similar, then the values of the activation energies of diffusion must also be very similar. This assumption is confirmed by the known experimental facts. The table shows the activation energy of self-diffusion and heterodiffusion of elements with similar physical properties (lattice type, atomic diameter, valence, etc.).

The results in the table apparently indicate that the activation energies of the diffusion of atoms of elements with similar properties in the same solvent are the same regardless of the type of crystal lattice of the solvent. It should be emphasized that this relationship also holds when the atomic diameters of elements are practically the same (analogs).

The role of the shape factor in diffusion proceeding by the vacancy mechanism becomes clear if one takes into account the fact that any difference between the size of the atoms of the impurity and the size of the principal atoms leads to the occurrence of elastic excitation of the lattice of the solvent around the impurity atom. The greater the difference between the diameters of the atoms of the substance in which the atoms diffuse, the stronger this effect, which influences the activation energy of the transfer of a foreign atom from a site to an intersite position (other properties of the diffusing atoms being the same).

The diffusion rate of tungsten in molybdenum and the self-diffusion rate of molybdenum are practically the same in the whole range of temperatures investigated, since in the equation mentioned not only the activation energies but also the preexponential factors are similar for self-diffusion of molybdenum ($D_0 = 4.52$ cm^2/sec) and for diffusion of tungsten in molybdenum ($D_0 = 3.18$ cm^2/sec).

According to Zener [5], the preexponential factor D_0 for metals with a cubic lattice can be expressed in terms of the entropy of activation ΔS, the interplane distance a, and the frequency of oscillation of the atoms in the lattice ν in the following way:

$$D_0 = a^2 \nu \exp\,(\Delta S/R) \qquad (4)$$

It was shown that the value of the entropy of activation of self-diffusion of pure metals with a cubic lattice determined from experimental values of D_0 is always positive. If we assume that in molybdenum $a = 2.72$ kX and $\nu = 10^{13} \cdot \sec^{-1}$ and neglect the linear expansion of the lattice, then the entropy of activation of self-diffusion of molybdenum calculated by (4) turns out to be equal to $+12.7$ cal/g–atom $\cdot$ deg, while it is $+12.0$ cal/g–atom $\cdot$ deg for the diffusion of tungsten.

Thus, according to Zener's theory, the entropy of activation of self-diffusion of molybdenum is greater than zero, and at the same time its value is close to that of the entropy of activation of diffusion of tungsten in molybdenum.

BOUNDARY DIFFUSION OF TUNGSTEN IN MOLYBDENUM

Since the physical properties of tungsten and molybdenum are very similar, we used W^{185} to study boundary diffusion in molybdenum.

The shape and size of the samples, the method of preparation and final treatment were described earlier in our discussion of the autoradiographic method of measuring the volume diffusion of tungsten in molybdenum.

It was shown by Fisher [6] that simultaneous volume and boundary diffusion in a polycrystalline sample leads to the following distribution of a diffusing substance deposited on the surface of the sample in a thin layer:

$$C = \exp\left[-\frac{y\,(2)^{1/2}}{(W)^{1/2}(\pi D_g\,\tau)^{1/4}\,(D_b/D_g)^{1/2}}\right]\mathrm{erfc}\,\frac{x}{2\sqrt{D_g\,\tau}} \,, \qquad (5)$$

where C is the concentration of the diffusing substance at a depth y and a distance x from the grain boundary considered to be a channel coinciding with the direction of the y axis. D_b and D_g are the coefficients of the boundary and volume diffusion, respectively; τ is the time of diffusional annealing; and W is the width of the boundary.

Since we are studying the only boundary diffusion ($x \approx 0$) by the autoradiographic method, we can simplify expression (5):

$$C = \exp\left[-\frac{y\,(2)^{1/2}}{(W)^{1/2}\,(\pi D_g\,\tau)^{1/4}\,(D_b/D_g)^{1/2}}\right]. \qquad (6)$$

If we replace the concentration of the diffusing substance C with the degree of darkening S of the autoradiogram at the corresponding point on the grain boundary and take the logarithm of (6), we obtain

$$\log S = \mathrm{const} - 0.461\,\frac{(D_b\,W/D_g)^{-1/2}}{(D_g\,\tau)^{1/4}}\,y. \qquad (7)$$

Equation (7) shows that under the conditions in which the solution given by Fisher is satisfied the values of log S must be a linear function of y and this makes it possible to calculate the coefficient of boundary diffusion from the tangent of the slope of the log $S = f(y)$ line when the coefficients of volume diffusion, the width of the boundary, and the time of diffusional annealing are known.

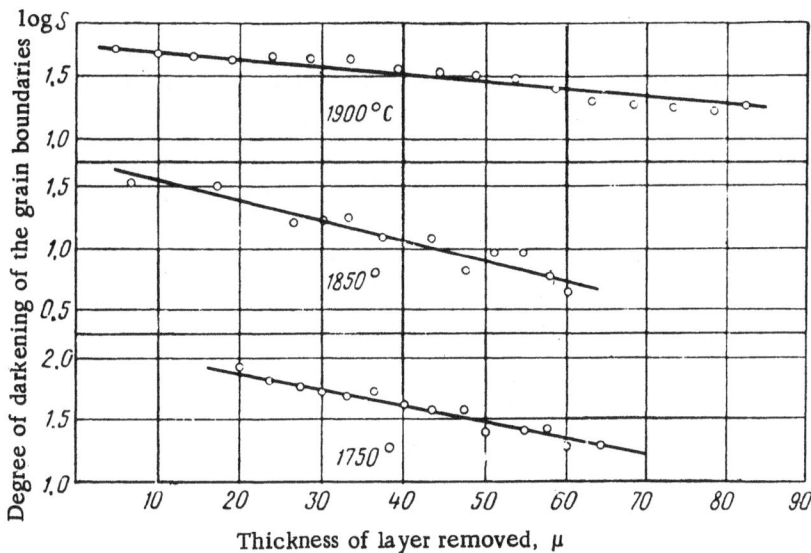

Fig. 4. Variation of the degree of darkening (S) of the grain boundaries on the autoradiogram with the thickness of the layer removed (diffusion of W^{185} in molybdenum).

Figure 4 shows the variation of the degree of darkening of grain boundaries on the autoradiogram with the depth of the diffusion layer in semilogarithmic coordinates for samples annealed 108 h at 1750°C, 103 h at 1850°C, and 99 h at 1900°C.

It can be seen that the degree of darkening (subtracting the background) is a linear function of the thickness of the diffusion layer at all temperatures. The coefficient of boundary diffusion was calculated by Eq. (7), assuming that the width of the boundary is 500 A [7]. The values of the coefficients of volume diffusion of tungsten in molybdenum necessary for these calculations were obtained in the first part of this work.

The following table shows the coefficients of boundary diffusion of tungsten in molybdenum.

Temperature of diffusional annealing, °C	1750°	1850°	1900°
Diffusion coefficients, D, cm²/sec	$5.4 \cdot 10^{-9}$	$1.3 \cdot 10^{-8}$	$2.0 \cdot 10^{-8}$

It should be noted that the coefficients of boundary diffusion of tungsten in molybdenum which were determined carry a constant error, since the real value of the width of the grain boundary is unknown. However, the data obtained allow us to calculate with sufficient precision the activation energy of boundary diffusion, since the constant error does not change the slope of the line representing the variation of log D_b with the reciprocal of the temperature. The temperature dependence of the coefficient of boundary diffusion of tungsten in molybdenum is very well represented by Eq. (8)

$$D_b = 1.1 \exp\left(-77000/RT\right) \text{ cm}^2/\text{sec.} \tag{8}$$

Remembering the implicit error in the width of the boundary, we can calculate the relative diffusional mobility of tungsten in molybdenum along the boundaries and within the grains. At 1750°C the rate of boundary diffusion exceeds the rate of volume diffusion 2100 times. At 1850 and 1900°C D_b/D_g is 1450 and 1200, respectively. These ratios decrease with increasing temperature because the activation energy of volume diffusion is greater than that of boundary diffusion.

The value of the activation energy of boundary diffusion obtained in this way is equal to 68% of the activation energy of diffusion in the lattice of the solvent. Such a ratio (0.61-0.68) has been found for many metals in the case of the vacancy mechanism of diffusion [8] and is confirmed in this work. Apparently this relationship is quite general; the idea of the strict ring mechanism of diffusion in molybdenum with a body-centered lattice becomes doubtful [9]. Let us note that the occurrence of diffusional porosities in the titanium-molybdenum system also contradicts the ring mechanism of diffusion [10].

According to the assumption made in [11], the elementary act of diffusion at the grain boundaries requires almost no energy for the formation of vacancies, because of the loosening of the structure in the inter-crystalline regions.

On the basis of these considerations, the energy of formation of vacancies Q_0 in molybdenum calculated from our results turns out to be 36 kcal/g–atom. For many metals the energy of the formation of vacancies varies from 0.3 to 0.4 of the activation energy of diffusion [8].

For molybdenum, $Q_0/Q_{dif} = 36/113 = 0.32$, which agrees satisfactorily with this relationship. It is interesting to note that a similar calculation for the self-diffusion of chromium [12] (belonging to the same group of the periodic table as molybdenum) gives a very similar value ($Q_0/Q_{dif} = 0.39$).

LITERATURE CITED

1. M. B. Bronfin, S. Z. Bokshtein, and A. A. Zhukhovitskii, Determination of the diffusion coefficients by the shift of the activity curve, Zavodskaya Lab. (7): 828-830, 1960.
2. M. B. Bronfin and N. N. Shabanov, Portable apparatus for removing parallel microlayers from metal samples, Zavodskaya Lab. (12): 1508-1510, 1962.
3. E. V. Borisov, P. L. Gruzin, L. V. Pavlinov, and G. B. Fedorov, Metallurgy and Metal Science of Pure Metals, Izd. MIFI, No. 1, 1959, pp. 213-218.
4. M. B. Bronfin, S. Z. Bokshtein, and S. T. Kishkin, Self-diffusion of molybdenum in molybdenum–zirconium alloys, in: Collection No. 3, Investigation of Alloys of Nonferrous Metals, Izd. Akad. Nauk SSSR, 1962, pp. 12-18.
5. C. Zener, Theory of D_0 for atomic diffusion in metals, J. Appl. Phys. 22(4): 372, 1951.
6. J. C. Fisher, Calculation of diffusion penetration curves for surface and grain boundary diffusion, J. Appl. Phys. 22 (1): 74, 1951.
7. V. I. Arkharov, Internal intercrystalline absorption in polycrystalline solids, Reports of the Institute of the Physics of Metals, UF, Akad. Nauk SSSR, No. 14, Sverdlovsk, 1954, p. 16.
8. S. D. Gertsriken and I. Ya. Dekhtyar, Diffusion of Metals and Alloys in the Solid State, Fizmatgiz, 1960, pp. 169-170.
9. A. D. Le Kler, Progress in Metal Physics, Metallurgizdat, 1956, p. 224.
10. P. G. Shewmon and I. H. Bechtold, Movements in titanium–molybdenum diffusion couples and the Zener theory of D_0, Acta Met. 3 (5): 452, 1955.
11. S. D. Gertsriken, Concerning the mechanism of diffusion, Fiz. Metal. i Metalloved. 2 (2): 378, 1956.
12. S. Z. Bokshtein, S. T. Kishkin, and L. M. Moroz, Autoradiographic investigation of self-diffusion and diffusion of chromium in some metals, Zavodskaya Lab. (3): 316, 1957.
13. R. E. Hoffman, P. W. Pikus, and R. A. Ward, Self-diffusion in solid nickel, J. Metals 8 (5) Sect. 2: 483-486, 1956.
14. R. C. Ruder and C. E. Birchenall, Cobalt self-diffusion, a study of the method of decrease in surface activity, J. Metals 191 (2): 142-146, 1951.
15. H. W. Mead and C. E. Birchenall, Diffusion of Co^{60} and Fe^{55} in cobalt, J. Metals 7 (9) Sect. 2: 994-995, 1955.
16. J. R. McEvan, J. U. McEvan, and L. Jaffe, Diffusion of Ni^{63} in iron, cobalt, nickel, and two iron–nickel alloys, Can. J. Chem. 37 (10): 1629, 1959.
17. G. S. Kuczynski, Measurement of self-diffusion of silver without using radioactive tracers, J. Appl. Phys. 21: 763, 1950.
18. R. A. Swalin, A model of solute diffusion in metals based on elasticity concepts, Acta Met. 5 (8): 443, 1957.

19. H. C. Gatos and A. D. Kurtz, Determination of the self-diffusion coefficients of gold by autoradiography, J. Metals 6 (5) Sect. 2:616-619, 1954.

20. W. Seith and A. Keil, Diffusion in Au—Pb and Ag—Pb legierungen, Z. Physik. Chem. 22 (5/6):350, 1933.

21. R. Resnick and L. S. C. Castleman, The self-diffusion of columbium, Trans. AIME 218 (2):307, 1960.

22. P. L. Gruzin and V. I. Meshkov, Self-diffusion of tantalum, in: Problems of Metal Science and the Physics of Metals, Collection No. 4, Metallurgizdat, 1955, p. 570.

23. B. I. Boltaks, Diffusion of impurities in germanium, Zh. Tekhn. Fiz. 26 (2):457, 1957.

24. Present article.

DIFFUSION OF RHENIUM IN MOLYBDENUM*

M. B. Bronfin

Alloys consisting of solid solutions of rhenium in molybdenum are very ductile at low temperatures and have a very high recrystallization threshold. The recrystallization temperature of molybdenum containing 50% Re is of the order of 1350°C. It can be rolled at room temperature to a thickness equal to 95% of the original thickness [1]. The limit solubility of rhenium in molybdenum is 50% (~35 at. %). Above this concentration the σ-phase precipitates, leading to embrittlement of the alloy. Dissolution and precipitation of phases in solids are to a great extent determined, as is well known, by the diffusional mobility of the atoms of the components of the alloy. To determine the time during which molybdenum alloys remain ductile at operating temperatures, at high temperatures, and during forming operations, one must know the diffusion constant of the alloy. To determine the heat resistance of molybdenum alloys one needs data on the diffusional mobility of atoms.

Knowledge of the thermal mobility of rhenium atoms in molybdenum is important from the practical as well as theoretical viewpoint, particularly if one takes into account the scarcity of data on diffusion in metals with high melting points.

Since we did not find any data in the literature on the diffusion of rhenium, we undertook measurements of the diffusional mobility of rhenium in metals, which is the subject of this first article.

MATERIALS AND METHOD

We used molybdenum smelted in a vacuum arc furnace. The ingots were rolled into bars 18 × 20 mm. Cylindrical samples 14 mm in diameter and 4-5 mm long were prepared from these bars. To decrease the contribution of boundary diffusion to the total diffusional flow of atoms, we annealed the samples in vacuum (10^{-3}-10^{-4} mm Hg) at 2000°C to obtain a large-grained structure.

We used Re^{186} as a radioactive tracer.

One end of the samples was electropolished and the samples were then activated in an electrolyte containing Re^{186} with an activity of the order of 10 mCi, the specific activity being 100-1000 mCi/g.

The radioactive electrolyte was prepared in the following way. Metallic Re^{186} was covered with a 10-15% aqueous solution of hydrogen peroxide and heated 10-12 h at a temperature below the boiling point until the decomposition of hydrogen peroxide was practically complete. Then 5 cc of concentrated sulfuric acid was added and the solution was kept at 60-70°C for 2 h. Distilled water was then added to the electrolyte to bring the total volume of the solution to 200 cc.

The conditions of electrolysis were as follows: current density, 1 A/cm²; temperature of the electrolyte 25-40°C; time, 30 min. The anode was platinum foil with an area several times the area of the activated sample. The radioactivity of the coating was 10,000-15,000 decompositions per minute. Control autoradiographs showed that the precipitation of Re^{186} over the entire surface of the sample is uniform under these conditions.

---

* N. N. Shabanov participated in the experimental part of this work.

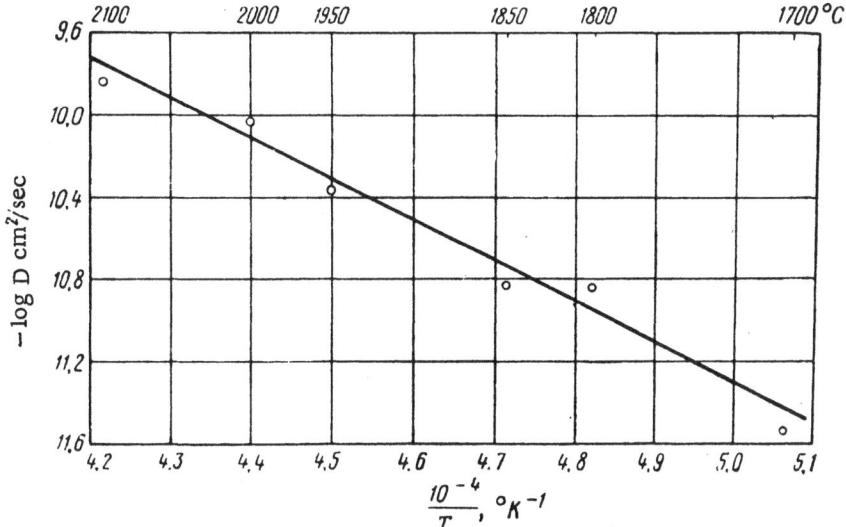

Temperature dependence of the diffusion coefficient of rhenium (Re[186]) in molybdenum.

The samples were subjected to diffusional annealing at 1700, 1800, 1850, 1950, 2000, and 2100°C in a TVV-2 vacuum resistance furnace. The diffusion coefficient of rhenium in molybdenum was determined by the shift in the activity curve [3]. The values of these coefficients are given below:

Temperature of diffusional annealing, °C	1700°	1800°	1850°	1950°	2000°	2100°
Diffusion coefficient, $D \cdot 10^{11}$ cm²/sec	0.280	1.30	1.40	4.70	8.40	17.0

RESULTS

The method of least squares applied to the results gives the following expression for the temperature dependence of the diffusion coefficient of rhenium in molybdenum:

$$D = 9.7 \cdot 10^{-2} \exp\left(-94700/RT\right) \text{ cm}^2/\text{sec.} \tag{1}$$

A corresponding graph drawn on the basis of the experimental points between 1700 and 2100°C is shown in the figure.

Elsewhere [4], we have derived the following equation for the self-diffusion of molybdenum:

$$D = 4.5 \exp\left(-113000/RT\right) \text{ cm}^2/\text{sec.} \tag{2}$$

According to this equation, the activation energy of self-diffusion of molybdenum is considerably higher than the activation energy of the diffusion of rhenium in molybdenum.

This latter result indicates that the diffusion rate of rhenium atoms is higher than the diffusion rate of the atoms of the solvent. Thus, the diffusion coefficient of rhenium in molybdenum is higher than the self-diffusion coefficient of molybdenum by a factor of 1.7 at 1900°C, a factor of 2 at 1800°C, and a factor of 2.5 at 1700°C. These facts agree with the assumption that the greater the difference between the physical and electrochemical properties of impurity atoms and the atoms of the solvent, the greater the rate of heterodiffusion.

The difference in the activation energies of diffusion of the atoms of the solvent and the impurity atoms, which in the present case reached 19 kcal/g-atom, is apparently due to the fact that the charges and sizes of the ions of the solvent and the impurity are different.

LITERATURE CITED

1. G. A. Geach and J. E. Hughes, The alloys of rhenium with molybdenum or with tungsten having good high-temperature properties, Plansee Proc. 1955, p. 245.
2. E. M. Savitskii, M. A. Tylkina, et al., Properties of rhenium—tungsten and rhenium—molybdenum alloys, in collection: Investigation of Heat-Resistant Alloys, Vol. 8, Izd. Akad. Nauk SSSR, 1962.
3. M. B. Bronfin, S. Z. Bokshtein, and A. A. Zhukhovitskii, Determination of the diffusion coefficient by the shift of the activity curve, Zavodskaya Lab. (7): 828-830, 1960.
4. M. B. Bronfin, S. Z. Bokshtein, and S. T. Kishkin, Self-diffusion of molybdenum, in molybdenum—zirconium alloys, in: Investigations of Alloys of Nonferrous Metals, Collection No. 3, Izd. Akad. Nauk, 1962, pp. 12-18.

INVESTIGATION OF THE STRUCTURE OF GRAIN BOUNDARIES
IN MOLYBDENUM AND ITS ALLOYS WITH ZIRCONIUM
AND RHENIUM BY THE INTERNAL FRICTION METHOD

S. Z. Bokshtein, M. B. Bronfin, S. T. Kishkin, and V. A. Marischev

The internal friction method, based on measurements of the damping of forced oscillations in polycrystalline samples, is a sensitive method of investigating the structure of a metal, including the structure of the grain boundaries.

In some metals the activation energy of the process of stress relaxation along the grain boundaries is close to the activation energy of self-diffusion, and as the result it has been assumed that the mechanism of boundary relaxation and the mechanism of volume diffusion are identical, and that at least in first approximation, the atomic structure of the grain boundaries cannot be very different from the structure of the grain itself [1].

Later investigations showed that the activation energy of boundary relaxation measured by the internal friction method is very different from the activation energy of self-diffusion [2], and as a rule is lower than the activation energy of self-diffusion.

It was found that impurities have a significant effect on the internal friction along the grain boundaries [3]. Winter and Weining [4] used the internal friction method to determine the degree of enrichment of the intercrystalline surfaces in the alloyed element in binary titanium alloys and found that the impurities increase the activation energy of intercrystalline relaxation and decrease the creep rate along the grain boundaries.

But the stress relaxation mechanism along the grain boundaries in pure metals, which is apparently changed by the presence of impurities, is not yet clear. It is quite possible that in this case the mechanism of viscous slip along the grain boundaries is to a certain extent replaced by diffusional transfer of the impurity atoms, which have a higher mobility than the atoms of the solvent.

Data on the internal friction along the grain boundaries in molybdenum are very interesting because the ductility of molybdenum is very sensitive to the structure of the grain boundaries.

EXPERIMENTAL METHOD

We measured the internal friction by torsional oscillations with a frequency of 0.3-0.4 Hz, using samples 100 mm long and ~0.6 mm in diameter.

The diagram of the apparatus used to measure the internal friction is shown in Fig. 1.* A vacuum of the order of 10^{-5}-10^{-6} mm Hg was created in the hermetically sealed container 12, using a low-vacuum pump 1, a diffusion pump 2, and a trap 8. The sample 11 was suspended from nichrome clamps 9 within the quartz tube 18, which passed vertically through the electrical furnace 19. A mirror 15 was attached to the lower part of the torsional pendulum at the intersection with a cross bar. To vary the natural frequency of oscillation of

---

* We modified an apparatus originally constructed by V. B. Osvenskii, using drawings provided by the Moscow Steel Institute.

the system, a weight 14 was added to the cross bar, using the sliding clamps 13. The sample was put in oscillation by turning the ferromagnetic cross bar with pulse electromagnets at the ends of the cross bar (not shown in Fig. 1). The bending oscillations were damped by damping the radial displacements of the vertical piece of the cross bar 16, which was immersed in vacuum oil at the bottom of the container 17. The damping of torsional oscillations of the sample was recorded visually by the deviation of the reflection of the light spot incident on the scale placed 5 m from the mirror 15. The temperature of the sample was measured with the thermocouple 10 introduced directly into the vacuum through the molybdenum glass stopper. The hot junction of the thermocouple was in the middle of the sample 2-3 mm from the axis of the sample. The cold junction was kept at 20°C. The emf of the thermocouple was measured with the PMS-48 potentiometer with a precision of ±0.001 mV, which corresponds to an error in temperature of the order of 1-1.5°C in the temperature range of 20-1000°C. The temperature in the furnace was kept constant within ±2°C either by hand or automatically.

The inner part of the furnace was connected to the gas generator consisting of the reaction tube 4, a trap 5, a MacLeod manometer 3, a container with a known volume 6, and a tube filled with activated charcoal 7. If necessary, known amounts of oxygen, nitrogen, or hydrogen (obtained by thermal decomposition of the corresponding compounds in the reaction apparatus) could be introduced into the furnace. At the same time, the humidity was frozen out and the components of the decomposed substances were precipitated by sublimation on the walls of the trap 5 placed in the gas supply line, which was cooled with liquid nitrogen or a mixture of dry ice and acetone.

In our experiments the torsion angle did not exceed 50'; the maximum tangential stress was about 0.5 kg/mm^2, which corresponds to a maximum deformation of the order of 10^{-5} on the surface. The axial load was 150 g ($\sigma = 0.5$ kg/mm^2). The logarithmic decrement divided by π constituted the measure of the internal friction. To decrease the effect of the surface state on the internal friction, all the samples were electropolished before being placed in the apparatus.

INTERNAL FRICTION IN UNALLOYED MOLYBDENUM

We measured the internal friction (Q^{-1}) in molybdenum 99.98% pure prepared by sintering. The samples were annealed 3 h at 1000°C in the apparatus, which produced a grain size not exceeding ~5 μ.

The temperature dependence of the internal friction in molybdenum is shown in Fig. 2. In molybdenum annealed at 1000°C the internal friction remains practically constant (0.001) up to a temperature of the order of 500°C. The curve begins to rise smoothly above this temperature. At temperatures between 600 and 700°C the $Q^{-1} = f(T)$ curve for the same sample rises sharply. This rise can apparently be related to the relaxation processes occurring at the grain boundaries.

For many metals we found a maximum internal friction along the grain boundaries at a temperature of about 0.4 of the melting point when the frequency of oscillation was close to 1 Hz. The temperature of the boundary peak (at a frequency of 1 Hz) is about 1000-1100°C.

The assumption that the increase of Q^{-1} between 600 and 700°C is due to the grain boundaries is confirmed not only by the fact that the temperature range of relaxation processes is the same as that of boundary internal friction but also by the dependence of the internal friction on temperature in the case of large-grained samples (see Fig. 2).

It is well known that when the structure of a metal approaches a monocrystalline structure, when the grain size becomes commensurable with the diameter of the sample, the peak of internal friction due to relaxation of stresses along the grain boundaries gradually decreases and disappears. The internal friction in molybdenum samples annealed 27 h at 2000°C is considerably lower than in samples annealed at 600-950°C. The internal friction of large-grained samples (3-4 grains in a diameter of 0.6 mm) at 800°C is one-fifth that in small-grained samples.

It should be noted that the curve representing the dependence of the internal friction on temperature (600-800°C) is reproducible after several heating cycles, and therefore cannot be attributed to irreversible recovery of recrystallization phenomena. The typical temperature dependence of the square of the frequency, which is proportional to the shear modulus, is shown in Fig. 2 for large-grained samples of molybdenum.

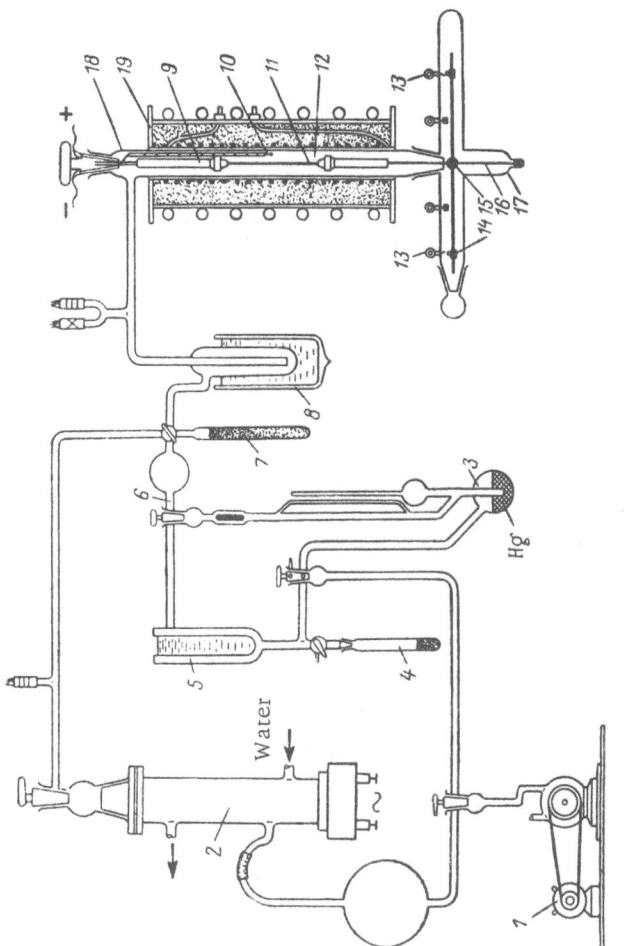

Fig. 1. Apparatus for measuring the internal friction in metals.

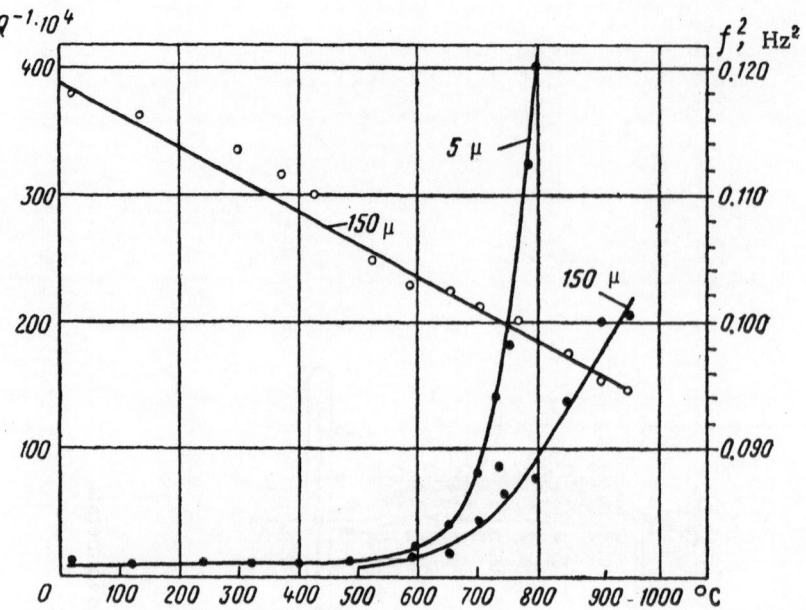

Fig. 2. Temperature dependence of the internal friction Q^{-1} and the square of the frequency of oscillation f^2 for pure molybdenum. ●) Internal friction; ○) square of the frequency of oscillation. The average diameters of the grains are indicated on the curves.

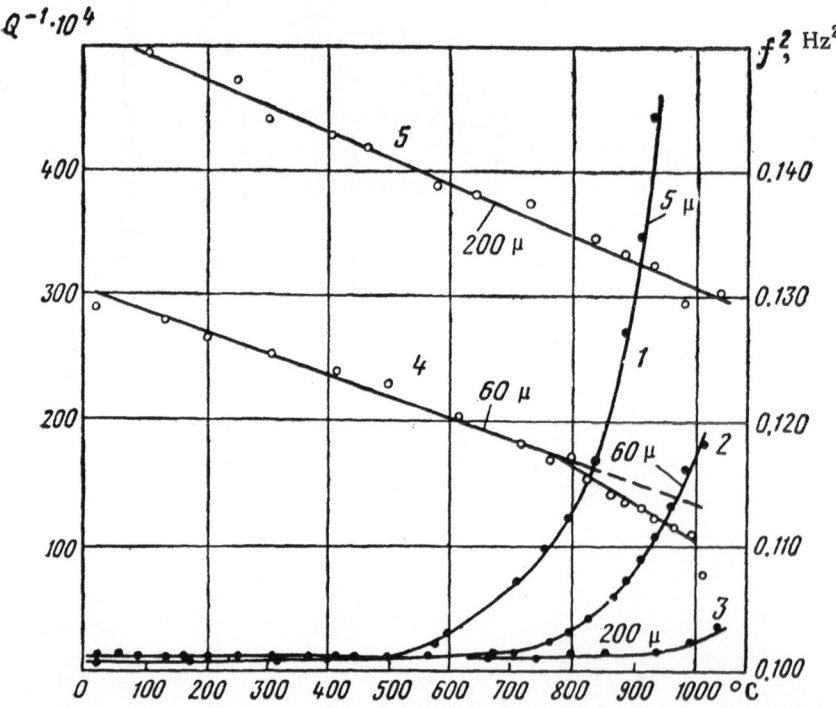

Fig. 3. Temperature dependence of the internal friction Q^{-1} and the square of the frequency of oscillation f^2 for Mo + 0.13% Zr. ●) Internal friction; ○) square of the frequency of oscillation. The average diameters of the grains are indicated on the curves.

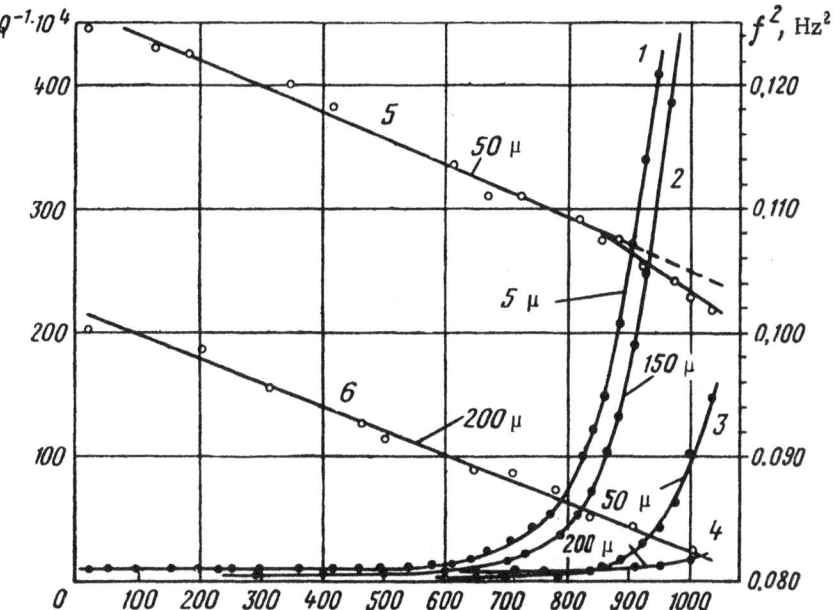

Fig. 4. Temperature dependence of the internal friction Q^{-1} and the square of the frequency f^2 for Mo+50% Re. ●) Internal friction; ○) square of the frequency of oscillation. The average diameters of the grains are indicated on the curves.

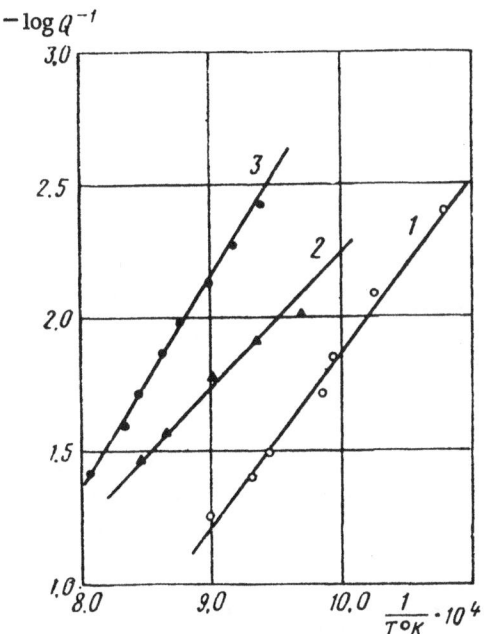

Fig. 5. Temperature dependence of the internal friction in the region of boundary relaxation. 1) Mo; 2) Mo+ 0.13% Zr; 3) Mo+50% Re.

In large-grained samples of molybdenum the shear modulus is a linear function of temperature, while in small-grained samples the dependence ceases to be linear (see Figs. 3 and 4). For small-grained molybdenum the degree of relaxation of the shear modulus at 800°C is of the order of 6%. This value is considerably lower than the total degree of relaxation, which has been found experimentally and derived theoretically for most metals (24-45%) [5].

These results indicate that under our experimental conditions the temperature of 800°C is relatively far from the temperature of the boundary friction peak. Therefore, we can determine the activation energy of internal friction at the boundary, which is usually determined by displacement of the temperature of the peak with changes in the frequency of oscillation.

It is well known that the internal friction Q^{-1}, which is related to the relaxation process (relaxation time τ), can be expressed in terms of the total degree of relaxation ΔM and rotational frequency of oscillation ω

$$Q^{-1} = \Delta M \frac{\omega \tau}{1 + (\omega \tau)^2} . \tag{1}$$

The temperature dependence of the relaxation time is

$$\tau = \tau_0 \exp\left(\frac{H}{RT}\right), \qquad (2)$$

where τ_0 is a constant of the material; H is the activation energy of the relaxation process; R is the gas constant; and T is the absolute temperature.

Equation (1) has a maximum at $\omega\tau = 1$. Since τ is an exponential function of temperature [see Eq.(2)], we can assume that $(\omega\tau)^2 \gg 1$ at temperatures relatively different from the temperature of the peak of internal friction if we take $\omega = 2\pi f$, where f is about 0.3 Hz. Therefore, with a precision which is sufficient for our case, Eq. (1) can be replaced with

$$Q^{-1} = \frac{\Delta M}{\omega} \cdot \frac{1}{\tau} = \frac{\Delta M}{\omega\tau_0} \cdot \exp\left[-\frac{H}{RT}\right], \qquad (3)$$

and, taking the logarithm of this expression, we obtain

$$\log Q^{-1} = \log\frac{\Delta M}{\omega\tau_0} - 0.4346\,\frac{H}{RT}. \qquad (4)$$

If we assume that $\log(\Delta M/\omega\tau_0) = \text{const}$ we can determine the activation energy H. Figure 5 shows that $\log Q^{-1}$ is a linear function of $1/T$.

In our experiments the activation energy of the internal friction of molybdenum along the grain boundaries was found to be 29.8 kcal/g-atom. This value is considerably lower than the activation energy of volume self-diffusion of molybdenum [6] (113 kcal/g-atom) and the activation energy of tungsten in molybdenum along the grain boundaries [7] (77 kcal/g-atom).

RESULTS—INTERNAL FRICTION IN MOLYBDENUM—ZIRCONIUM AND MOLYBDENUM—RHENIUM ALLOYS

The molybdenum—zirconium alloys investigated contained 0.13% Zr, 0.008% C, 0.006% O_2, and 0.007% H_2. The internal friction was measured after annealing at 1000, 1850, and 2000°C. The temperature dependences of internal friction and the square of the frequency of oscillation for molybdenum—zirconium samples are shown in Fig. 3. With increasing annealing temperatures the grain size increases, the internal friction along the boundary measured at the same temperature decreases, and the degree of relaxation of the shear modulus decreases from 8% at 900°C (the initial state of the sample) almost to zero after annealing 27 h at 2000°C (the average diameter of the grains being about 200 μ). For the molybdenum—zirconium alloy the activation energy of the relaxation process along the grain boundaries in the initial state (i.e., the average diameter of the grains was 5 μ) is 23.2 kcal/g-atom, while after annealing 36 h at 1850°C (average diameter of the grains about 60 μ) it is 22.4 kcal/g-atom. Thus, the activation energy of the internal friction along the grain boundaries in the molybdenum alloy containing 0.13% Zr remains almost constant, although the grain size increases from 5 to 60 μ. This low degree of internal friction (of the order of 0.001) resulting from annealing at 2000°C remains about the same down to 900°C; after that the $Q^{-1} = f(T)$ curve rises very slightly (see Fig. 3).

Figure 4 shows a family of $Q^{-1} = f(T)$ and $f^2 = \varphi(T)$ curves for the molybdenum alloy containing 50% Re.* The alloy was deformed by drawing at room temperature (the decrease of the cross section being 60%). The temperature dependence of the internal friction of the molybdenum—rhenium sample in the initial state remains practically constant up to 700°C (curve 1). Q^{-1} begins to increase at 750-800°C. The reheated drawn samples give the same internal friction curve, but it is shifted 30° (curve 2) toward higher temperatures. After recrystallization for 5 h at 1500°C (the average diameter of the grains being about 50 μ) the internal friction in the region of boundary relaxation decreases and the degree of relaxation of the shear modulus decreases

_____

* Our thanks to E. M. Savitskii and M. A Tylkina for providing the samples of this alloy.

from 3% in the initial state (900°C) to 0.4%. The molybdenum alloy containing 50% Re was annealed 33 h at 1950°C (the average diameter of the grains being 200 μ). The internal friction of this alloy is very low (about 0.001) and remains constant between 20 and 1000°C. No relaxation of the shear modulus was observed.

The activation energy of internal friction along the boundaries in the molybdenum—rhenium sample is 35 kcal/g-atom in the initial state and 44.3 kcal/g-atom after recrystallization at 1500°C. In this case the slight increase in the activation energy is apparently due to the transformation of the metal from the deformed state into a state with a higher degree of equilibrium as the result of recrystallization annealing. Both values of the activation energy of the internal friction of the molybdenum alloy containing 50% Re at high temperature are several times lower than the activation energy of the self-diffusion of molybdenum in the alloy containing 20% Re (139 kcal/g-atom) and lower than the activation energy of diffusion of rhenium in molybdenum (94.7 kcal/g-atom).*

DISCUSSION OF RESULTS

The values of the internal friction of molybdenum and its alloys with zirconium and rhenium show that boundary relaxation increases at different temperatures in the different materials. For example, let us compare the temperature dependences of the internal friction and the square of the frequency of oscillation of these materials in the initial state (grain size about 5 μ).

If we assume that internal friction begins to increase at the temperature at which the $Q^{-1} = f(T)$ curve ceases to be linear, then this temperature for the molybdenum—rhenium alloy turns out to be 700°C, while for unalloyed molybdenum and for molybdenum containing 0.13% Zr it is about 600°C. Correspondingly, the curve of internal friction of molybdenum containing 50% Re is shifted much further toward higher temperatures. Beginning at 700°C, unalloyed molybdenum has the highest degree of internal friction, while the alloy with 50% Re has the lowest. The absolute values of the internal friction at 800°C (the average diameter of the grains being 5 μ and the frequency of oscillation 0.3-0.4 Hz at 20°C) are 0.04 for Mo, 0.0126 for Mo+0.13% Zr, and 0.04 for Mo, 0.0126 for Mo+0.13% Zr, and 0.004 for Mo+50% Re. However, it does not follow that the effect of impurities on the value of the internal friction along the boundaries is due only to the impeding of the elementary acts of stress relaxation on the grain boundary.

Figure 5 shows the temperature dependences of Q^{-1} for different materials in the initial state (the average diameter of the grains being about 5 μ). If the value of the internal friction along the grain boundaries were determined only by the activation energy then it should be highest in the Mo—Zr alloys and not in pure Mo, which was found experimentally. One of the possible explanations of the results shown in Fig. 5 is that in spite of the decrease in the activation energy resulting from the addition of small amounts of zirconium to molybdenum, the number of sites where the probability of relaxation is high also decreases relatively sharply.

The relative values of the activation energy of boundary relaxation for different materials indicate that the process responsible for boundary relaxation can hardly be the diffusional displacement of substitution atoms.

It seems more reasonable to assume that the mechanism of boundary relaxation is related to the migration of impurities which penetrate the intercrystalline zone as the result of the temperature and cyclic stress. The regrouping of oxygen, carbon, nitrogen, and other impurity atoms in the field of stresses (e.g., ascending diffusion) requires a much lower expenditure of energy than the transition of the more inert atoms (from the diffusional standpoint) which normally occupy the lattice sites. This circumstance apparently explains the relatively low value of the activation energy of the boundary internal friction found in this investigation.

Qualitatively, the effect of substitution impurities can be explained in terms of their interaction with migrating atoms of penetration impurities and also by their capacity to change, to some degree, the degree of structural imperfection of the intercrystalline zone.

The elements which have great affinity for penetration atoms can apparently decrease their diffusional mobility and also (under certain conditions) decrease their concentration in the intercrystalline zone, which

---

*See the article by Bronfin on the diffusion of rhenium in molybdenum in this collection, p. 24.

leads to a decrease of the internal friction along the grain boundaries. The more perfect structure of the inter-crystalline zone and the decrease of its free energy as the result of internal absorption of affinity impurities [8] also decrease the scattering of oscillation energy recorded as internal friction along the grain boundaries.

The conditions of diffusion of penetration atoms under the effect of temperature and applied stress are apparently changed in a rather complex manner when Zr and Re are present in the intercrystalline zone of Mo. Zirconium, which has a great affinity for oxygen and carbon, should "prevent" a certain number of atoms of impurities from participating in diffusional transitions, and thus decrease the internal friction.

On the other hand, since the sizes of zirconium and molybdenum are very different, additional excitation of the lattice must be created around the sites in which zirconium is substituted for molybdenum, and this additional lattice excitation is equivalent to a decrease in the effective activation energy of "unbound" atoms of penetration impurities. This latter effect is apparently translated into some decrease of the activation energy of boundary internal friction in molybdenum−zirconium alloys as compared to the activation energy in un-alloyed molybdenum.

The addition of 50% Re to molybdenum considerably increases the activation energy of internal friction along the grain boundaries as the result of the "improvement" of the intercrystalline zones of the metal and as a result of the decrease of the concentration of penetration atoms in these zones due to the increased solu-bility of these impurities within the grain.

Since there is a definite relationship between the creep processes and the process of stress relaxation, one may assume that, other conditions being equal, the molybdenum alloy containing 50% Re has the highest resistance to creep along the grain boundaries. This alloy had the highest activation energy of boundary re-laxation among the materials investigated.

LITERATURE CITED

1. Ke Tin-Sui, Stress relaxation along the grain boundaries in metals, in collection: Elasticity and In-elasticity in Metals [Russian translation], (S. V. Vonsovskii, ed.), IL, 1954, p. 234.
2. L. Rotherham and S. Pearson, Internal friction and grain boundary viscosity of copper and binary copper solid solutions, J. Metals 8 (8) Sec. 2: 881, 1956; S. Pearson and L. Rotherham, Internal friction and bound-ary viscosity of silver and binary silver solid solutions, ibid.
3. Ke Tin-Sui, Model of grain boundaries and mechanisms of viscous intercrystalline slip, in collection: Elasticity and Inelasticity in Metals [Russian translation], (S. V. Vonsovskii, ed.), IL, 1954, p. 313; O. I. Datsko, Variation of the properties of grain boundaries in nickel as the result of alloying with copper, in: Reports of the Institute of the Physics of Metals, UF Akad. Nauk SSSR, No. 22, Sverdlovsk, 1959, p. 117.
4. I. Winter and S. Weining, The grain boundary adsorption of solutes, Trans. AIME 215 (1): 74, 1959.
5. Ke Tin-Sui, Inelastic properties of iron, in collection: Elasticity and Inelasticity in Metals [Russian translation], (S. V. Vonsovskii, ed.), IL, 1954, p. 300.
6. M. B. Bronfin, S. Z. Bokshtein, and S. T. Kishkin, Self-diffusion of molybdenum in molybdenum−zir-conium alloys, in collection: Investigation of Alloys of Nonferrous Metals, No. 3, Izd. Akad. Nauk SSSR, 1962, p. 12.
7. S. Z. Bokshtein, M. B. Bronfin, and S. T. Kishkin, Volume and boundary diffusion of tungsten in mo-lybdenum, this volume, p. 16.
8. V. I. Arkharov, Intercrystalline adsorption in solid solution, in: Reports of the Institute of Metallurgy, UF Akad. Nauk SSSR, No. 20, 1956, p. 16.

INTERNAL FRICTION IN DEFORMED MOLYBDENUM ALLOYS

M. B. Bronfin and V. A. Marischev

The curve representing the temperature dependence of the internal friction in cold deformed metals sometimes has a peak which decreases after low-temperature annealing and disappears completely after recrystallization of the alloy. For iron, for example, such a peak occurs at 200°C at a frequency of oscillation of 1 Hz [1].

Investigation of the internal friction in deformed iron containing nitrogen and carbon [2] showed that the peak of internal friction increases with the degree of deformation. The peak is displaced toward low temperatures. The maximum height of this deformation peak, which occurs only when there are dissolved impurities, depends on the amount of impurities. It decreases considerably when nitrides and carbides precipitate from the solid solution.

Internal friction can be increased only by relatively free dislocations in the subgrains [3].

During the movement of dislocations, a stress is created between the dislocations and the impurities forming Cottrell clouds. At temperatures at which diffusion can occur these slowly moving dislocations will carry the impurity clouds with them [4].

The deformation peak in iron at 200°C was explained by a mechanism based on all these assumptions [2].

Cyclic shear stresses of a frequency of 1 Hz are insufficient to detach these dislocations completely from the impurity cloud. However, they induce oscillations of dislocations around the impurities which are not accompanied by any considerable scattering of energy. At low temperatures the internal friction does not increase, since the impurity atoms cannot follow dislocations (adiabatic case). The diffusion of impurities is facilitated at high temperatures, and therefore the motions of dislocations and impurity atoms are in phase at any given moment (isothermal case). Between these two limit cases the oscillations of the impurity atoms lag behind the oscillations of the dislocations. With increasing temperature the phase difference between these two oscillations increases to a maximum value and the maximum internal friction in iron at 200°C corresponds to this value.

The present article concerns an investigation of the internal friction in deformed molybdenum and its alloys. We determined the dependence of the temperature of the maximum internal friction resulting from plastic deformation on the grain size.

We investigated sintered molybdenum and cast binary alloys of molybdenum containing 0.13% zirconium and 50% rhenium.

METHOD AND RESULTS

Samples 110-120 mm long were cut from wire deformed by drawing. The experiments were made with an apparatus for measuring internal friction by the method of torsional oscillations. To avoid oxidation during the experiments, the apparatus was evacuated to 10^{-5} mm Hg. The temperature was kept constant with a precision of 1°. The average heating rate between 20 and 1000°C was 2°/min. The maximum deformation did not exceed 10^{-5}. The frequency of torsional oscillation during measurements of internal friction was 0.35-0.40 Hz.

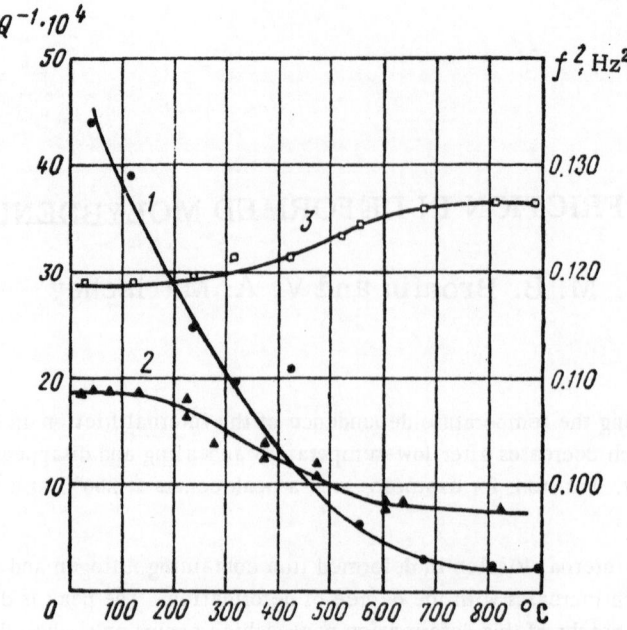

Fig. 1. Variation of the internal friction and the square of the frequency of oscillation of Mo + 0.13% Zr at 20°C with the annealing temperature. 1) Internal friction of the samples whose surface layers were removed mechanically; 2) the same for electropolished samples; 3) variation of the square of the frequency of oscillation for the same sample.

It was found that grain growth in the surface layers of the wire samples was greatly impeded during recrystallization annealing.

Therefore, we removed the outer layer of the wire to make the structure approximately uniform through the entire section of the samples. Since mechanical removal of the outer layer leads to a great increase in internal friction at 20°C (Fig. 1), the samples were electropolished in an alcohol solution of sulfuric acid. The original diameter of the samples was 0.9 mm; it was reduced to 0.5-0.6 mm after electropolishing.

Figures 2, 3, and 4 in the preceding article show the temperature dependence of the internal friction and the square of the frequency of oscillation (a value proportional to the shear modulus) of molybdenum and its alloys with zirconium (0.13% Zr) and rhenium (50% Re). The sharp increase in the internal friction at temperatures of 700-1000°C and also a significant drop of the shear modulus are the result of relaxation processes along the grain boundaries [5, 6]. In [5] no relaxation of the modulus was observed in this temperature range.

The curves representing the temperature dependence of Q^{-1} in molybdenum and its alloy with zirconium deformed at room temperature show not only an increase in the internal friction due to the effect of the boundary but also a peak at 200-300°C which completely disappears after 3 h at 900°C. *

Figure 2 shows that after repeated deformation the maximum internal friction is somewhat lower, while the temperature at which the peak occurs as the result of plastic deformation remains almost the same.

It should be emphasized that the curves shown in Fig. 2 were obtained at the same frequency of oscillation, which makes it possible to compare them. The frequency was maintained constant by leaving the samples in the clamps of the apparatus during bending.

---

* Since we examined only the qualitative relationship in this investigation, we did not measure the plastic deformation precisely.

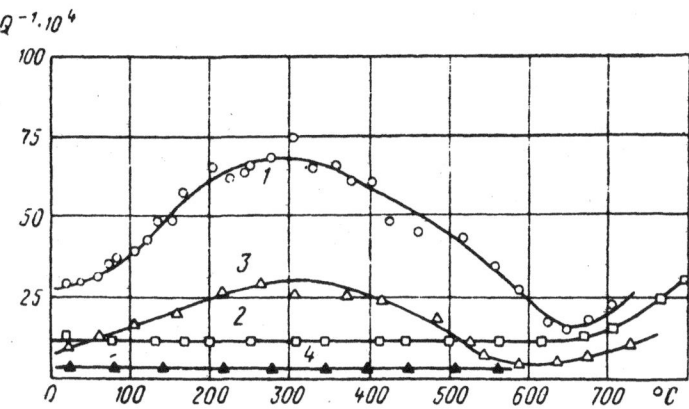

Fig. 2. Internal friction of Mo + 0.13% Zr recrystallized at 1850°C. 1) Sample after plastic deformation; 2) same sample after 3 h annealing at 900°C; 3) same sample after repeated deformation; 4) same sample after annealing 3 h at 900°C.

To determine the effect of grain size on the deformation peaks of internal friction in molybdenum and its alloys with zirconium and rhenium, each sample was recrystallized at three different temperatures. Figure 3 shows that the maximum internal friction due to plastic deformation of the recrystallized molybdenum—zirconium alloy shifts toward higher temperatures with increasing grain size.

No peak of internal friction was observed for the molybdenum—rhenium alloy between room temperature and 700°C. The variation of the internal friction with temperature for this alloy is shown in Fig. 3, where it can be seen that the curves for deformed and annealed samples coincide.

The internal friction at 20°C decreases continuously with increasing annealing temperatures, and the rate of increase is particularly high between 200 and 600°C. The internal friction measured at 20°C remains almost constant in samples kept at temperatures above 600°C. The value of the square of the frequency of oscillation of the sample (proportional to the shear modulus) increases most in the same temperature range in which the internal friction decreases.

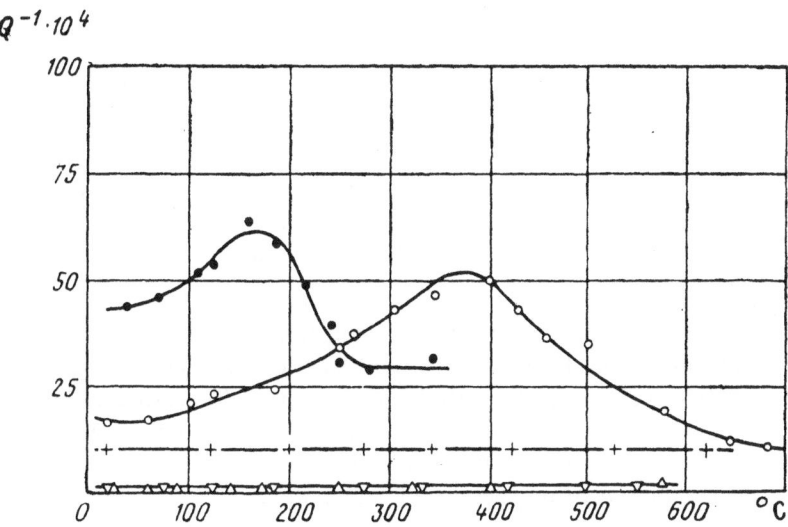

Fig. 3. Temperature dependence of the internal friction of deformed molybdenum alloys. ●) Mo + 0.13% Zr recrystallized at 1600°C; ○) same alloy recrystallized at 2000°C; −+−+) same alloy recrystallized at 2000°C, deformed, and then annealed 3 h at 900°C; Δ) deformed Mo + 50% Re; ∇) same alloy after annealing 3 h at 900°C.

DISCUSSION OF RESULTS

According to the modern theory, the dislocational structure of a crystal can be represented as a three-dimensional network whose segments, fixed at the sites, can bend under the effect of very low stresses. The total deformation E_{tot} in the elastic region resulting from stress is the sum of the deformation E_{at} resulting from the displacement of atoms and the deformation E_d resulting from bending of dislocation loops.

$$E_{tot} = E_{at} + E_d. \tag{1}$$

By dividing all the terms in Eq. (1) by the stress, we obtain

$$\frac{1}{G_d} = \frac{1}{G_0} + \frac{E_d}{\sigma}, \tag{2}$$

$$\frac{G_0 - G_d}{G_0} = \frac{\Delta G}{G_0} \approx E_d G_0/\sigma, \tag{3}$$

where G_0 is the shear modulus of the material not containing dislocations and G_d is the shear modulus of material containing dislocations.

Equation (3) shows that the material containing dislocations has a lower shear modulus than the material without dislocations. Plastic deformation increases the dislocation density within the grains. This leads to a decrease of the effective shear modulus and to an increase of the internal friction.

The subgrains of freshly deformed metal contain considerable numbers of deformations [7]. These internal deformations decrease as the result of polygonization during annealing at temperatures below the recrystallization temperature. Since the density of relatively free dislocations in the subgrains decreases, the value of E_d in Eq. (3) decreases, and as a result the shear modulus approaches its normal value.

From this viewpoint, the curves in Fig. 1 would indicate that intense pinning of dislocations as the result of polygonization, interaction with point defects, and other processes occurs at temperatures between 250 and 600°C. Low-temperature peaks of internal friction occur in the material investigated only after plastic deformation. The high-temperature branches of these peaks are located approximately in the same temperature range (Fig. 3) in which the internal friction and the defect of the shear modulus (measured at 20°C–Fig. 1) decrease rapidly. Apparently, these results confirm the dislocational nature of the processes inducing deformational peaks.

To interpret the data concerning the temperature dependence of internal friction at temperatures between room temperature and 700°C, we shall use the assumptions made in [2]. Molybdenum and α-iron have the same lattice, and therefore the interaction between dislocations and the impurity atoms in both materials is qualitatively the same.

The displacement of the deformational peak of internal friction toward lower temperatures with increasing degrees of plastic deformation[2], as well as the shift of a similar peak in the opposite direction with increasing grain size can apparently be explained in the following terms.

The peak of internal friction due to plastic deformation occurs only for a given relationship between the mobility of the oscillating dislocations and the penetrated atoms of the impurity. In first approximation, the rate of diffusion of impurity atoms in the vicinity of dislocations is independent of the degree of plastic deformation, and the effective velocity of the oscillational motion of dislocations must decrease with increasing cold hardening. It follows, then, that the temperature of the maximum phase difference between these displacements which characterizes the ratio between the rates of displacement of dislocations in atoms forming Cottrell clouds decreases with increasing degrees of deformation.

The shift of the deformation peak of internal friction toward higher temperatures (see Fig. 3) for large-grained samples can be explained, according to[2], if one assumes that at a given degree of deformation the density of relatively free dislocations affecting the internal friction decreases with increasing grain size. If the motion of dislocations in large-grained samples is easier than in small-grained samples, then at higher temperatures the ratio between the mobilities of dislocations and impurities ensuring the maximum phase difference between the motion of dislocations and the penetrated atoms, and consequently the peak of internal friction, will shift toward higher temperatures.

The irreversible decrease of the internal friction at temperatures higher than the temperature of the peak is due, in our opinion, not only to the increase in the diffusional mobility of impurity atoms but also to the decrease in the number of relatively free dislocations as the result of polygonization.

As we have noted before, it is at these temperatures that one observes an intense decrease of internal friction and an increase of the effective shear modulus measured at 20°C.

The fact that the curve representing the variation of internal friction with temperature does not have a peak in the case of the molybdenum—rhenium alloy leads us to conclude that the mechanisms of deformation of this alloy, the molybdenum—zirconium alloy, and pure molybdenum are different. One may assume that because of the additional deformation mechanism—twinning—characteristic of the molybdenum alloy containing 50% Re, plastic deformation of this alloy is not accompanied by the number of relatively free dislocations necessary for the occurrence of the deformation peak of internal friction, as is the case in pure molybdenum and the molybdenum—zirconium alloy.

LITERATURE CITED

1. I. Snoek, Effect of small quantities of carbon and nitrogen on the elastic and plastic properties of iron, Physics 8: 711-733, 1941.
2. W. Köster, L. Bangert, and R. Hahn, Das Dampferungsverhalten von gerecktem technischem Eisen, Arch. Eisenhuettenw. 25 (11/12): 569-578, 1954.
3. A. S. Novik, Recovery of internal friction and the elastic constants, in collection: Creep and Relaxation, Metallurgizdat, 1961; Internal friction in metals, Usp. Fiz. Metal., Vol. 1, 1956.
4. G. Shoek, Theory of creep, in collection: Creep and Relaxation, Metallurgizdat, 1961.
5. I. L. Mirkin, V. Z. Tseitlin, and G. G. Morozova, Internal Friction and Shear Modulus in Some Pure Metals Composing Heat-Resistant Alloys, TsNIITMASh Reports, Book 101, Mashgiz, 1961.
6. This volume, p. 27.
7. A. S. Novik, Recovery of internal friction and the elastic constants, in collection: Creep and Relaxation, Metallurgizadat, 1961.

EFFECT OF ANNEALING ON DIFFUSION
FOLLOWING PLASTIC DEFORMATION

S. Z. Bokshtein, T. I. Gudkova, A. A. Zhukhovitskii, and S. T. Kishkin

We have shown previously [1] that stress and deformation accelerate diffusion processes in metals and complex alloys.

The acceleration of diffusion resulting from plastic deformation can be related to changes in the state of the metal, particularly changes in the microstructure, which are reversible, difficult to reverse, or irreversible.

Experimental results on the effect of plastic deformation at high temperature and room temperature on the diffusion rates of tin in nickel (Tables 1 and 2) confirm this conclusion.

Tables 1 and 2 show that plastic deformation of nickel at room and high temperatures accelerates diffusion considerably during subsequent annealing at 800°C. The effect of deformation at high temperature is much smaller than at low temperature, although deformation at high temperature still has a great effect on the diffusion of tin in nickel.

We measured the effect of high-temperature annealing on the stability of structural changes induced by plastic deformation.

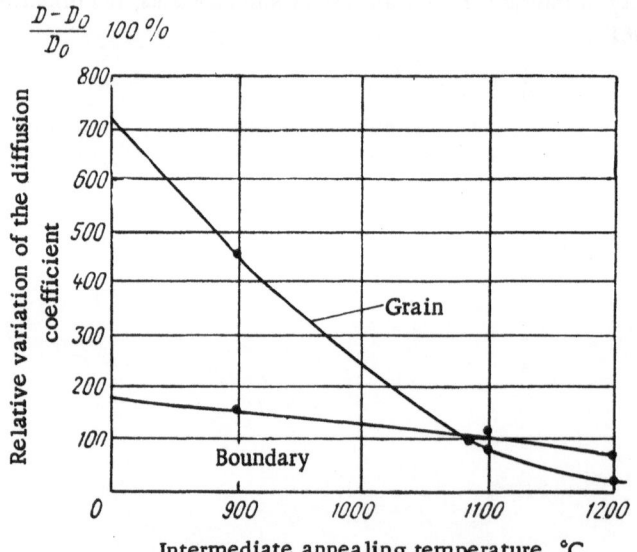

Effect of intermediate annealing on the diffusion of tin in nickel
(deformed at 700°C, σ = 6 kg/mm², δ = 12.1%).

TABLE 1. Effect of Plastic Deformation at Room Temperature on the Diffusion of Tin in Nickel at 800°C

Deformation, %	Through the grain		Along the boundary	
	$D \cdot 10^{13}$ cm^2/sec	$\dfrac{D-D_0}{D_0} \cdot 100\%$	$D \cdot 10^{12}$ cm^2/sec	$\dfrac{D-D_0}{D_0} \cdot 100\%$
0	5.2	—	8.3	—
5	22.0	325	14.0	70
10	46.0	800	20.0	140

TABLE 2. Effect of Plastic Deformation at 700°C on the Diffusion of Tin in Nickel at 800°C

Deformation, %	Through the grain		Along the boundary	
	$D \cdot 10^{13}$ cm^2/sec	$\dfrac{D-D_0}{D_0} \cdot 100\%$	$D \cdot 10^{12}$ cm^2/sec	$\dfrac{D-D_0}{D_0} \cdot 100\%$
0	4.8	—	7.6	—
4.7	9.6	100	13.0	71
10.0	24.2	400	18.1	135
13.0	43.0	800	22.0	200

TABLE 3. Effect of Plastic Deformation and Intermediate Annealing on the Diffusion of Tin in Nickel at 800°C

State of the material	Through the grain		Along the boundary	
	$D \cdot 10^{13}$ cm^2/sec	$\dfrac{D-D_0}{D_0} \cdot 100\%$	$D \cdot 10^{12}$ cm^2/sec	$\dfrac{D-D_0}{D_0} \cdot 100\%$
Initial	5.2	—	7.5	—
Deformed 12.1%	42.6	720	21.0	180
Deformed and annealed				
at 900°C	29.1	460	18.7	150
at 1100°C	9.4	80	15.8	110
at 1200°C	6.2	20	12.0	60

The samples were deformed at 700°C under a stress of 6 kg/mm^2 and then annealed at 900, 1100, and 1200°C for 25 h. The samples were then subjected to diffusional annealing at 800°C for 125 h. The diffusion coefficients of tin in nickel with and without intermediate annealing are shown in Table 3.

From an analysis of these data we conclude that intermediate annealing weakens the effect of deformation on diffusion, but does not completely eliminate it (see the figure).

The fact that intermediate annealing has a much smaller effect on boundary diffusion than on volume diffusion can be explained either by the fact that deformation affects essentially the volume of the grain or by the fact that changes in the boundaries induced by plastic deformation are so great (microcracks are formed) that subsequent annealing cannot eliminate them.

Apparently plastic deformation at high temperature or room temperature leads to irreversible changes in the microstructure of the alloy or changes that can be eliminated only by heating to a very high temperature or for a long time.

It is well known that plastic deformation at low as well as at high temperatures leads to polygonization of the structure. The data obtained in [2] are very interesting in this respect. In this investigation nickel and nickel—titanium alloys were subjected to 1.0, 2.3, and 6.0% elongation and then annealed at 800°C. Microscopic and x-ray examinations showed that the grains remain almost unchanged, but blocks are formed in the crystals. The number of these blocks increases with the degree of deformation.

Thus, a degree of deformation relatively low as compared to those used in our investigation produces a block structure which is not eliminated by subsequent annealing at 800°C.

It is well known that plastic deformation decreases the size of blocks and increases their degree of disorientation in a number of metals. This effect is particularly great during the initial stages of plastic deformation [3].

X-ray analyses of nickel samples subjected to plastic deformation at 700°C (the deformation being 15.6%) and subsequent diffusional annealing for 135 h at 800°C showed that the number of fragmented particles increases five to six times as compared to the number in the initial state.

The fragmentation of particles and the development of block structure accelerate diffusion.

Thus, although intermediate annealing decreases the effect of plastic deformation, which accelerates diffusion, this effect is completely eliminated only after annealing at a very high temperature—a temperature considerably higher than the recrystallization temperature.

LITERATURE CITED

1. S. Z. Bokshtein, T. I. Gudkova, A. A. Zhukhovitskii, and S. T. Kishkin, Some Problems of the Strength of Metals, Izd. Akad. Nauk SSSR, 1959, p. 76; B. S. Bokshtein and T. I. Gudkova, Izv. Vysshikh. Uchebn. Zavedenii, Ser. Chernaya Met. 5: 108, 1960; F. S. Buffington and M. Cohen, J. Metals 4 (8): 859, 1952.
2. E. Parker and T. Hazlitt, Structure and Properties of Metals [Russian translation], Metallurgizdat, 1957.
3. L. S. Moroz, Fine Structure and Strength of Steels, Metallurgizdat, 1957.

EFFECT OF THE SURFACE STATE ON SELF-DIFFUSION
AND DIFFUSION IN SURFACE LAYERS OF ALLOYS

S. Z. Bokshtein, M. A. Gubareva, and S. T. Kishkin

The surface state of a metal has, in a number of cases, a decisive effect on its behavior during use. It is well known that destruction of a metal usually begins at the surface. The various mechanical and heat treatments and the medium affect the surface state of the metal considerably, i.e., the structure and the composition of the surface layers change.

It is always extremely difficult to determine the state of the metal in the surface layer because the surface layer is so thin.

Polishing of copper and iron produces plastic deformation of the surface layer and the formation of dispersed mutually disoriented fragments [1].

X-ray analysis of a zinc single crystal showed that, after polishing, the surface of the sample consists of small fragments with very different orientations [2].

Polishing with abrasives produces a great number of scratches. A study of the scratched surfaces of lithium fluoride crystals showed that dislocations occur around scratches and that the volume of these dislocations is 100 times greater than the volume of the scratch [3, 4]. This region around the scratch has a higher solubility [5] and the density of dislocations decreases with increasing distance from the scratch.

The dislocation density in the surface layer of a polished crystal of lithium fluoride is 10^8-10^9 cm^{-2}[6].

A number of investigators have studied the thermal stability of deformations resulting from polishing. For example, in [7] the author studied the surface of a zinc single crystal after polishing and annealing. He showed that annealing does not restore monocrystallinity, but produces recrystallization of the surface layer. The grain size increases with the distance from the surface of the single crystal and the thickness of the recrystallized layer increases with the annealing temperature.

Data on the thermal stability of deformations next to isolated surface defects (scratches and pits) were obtained in [8]. These authors showed that in this region of the crystal the degree of stability of dislocations increases with the density of dislocations, and in the immediate vicinity of the defects the dislocations are stable up to a very high temperature.

The investigation of the removal of defects in polycrystalline copper induced by polishing [9] showed that the defects in the surface layer are stable at high temperatures.

According to [10], the possible cause of the stability of the deformed thin surface layer is the partial separation of the elements of the dispersed structure. This separation can be caused either by the condensation of excess vacancies and dislocation lines on the boundaries of the structural elements or by the thermal decomposition of thin oxide layers situated along the boundaries of the structural elements.

It is well known that grinding and polishing are accompanied by partial oxidation of the metal. It is logical to assume that the surface states of metals must have significant effects on the properties and the kinetics of various processes, particularly the kinetics of diffusion in the surface layer.

TABLE 1. Self-Diffusion Coefficients of Nickel in the Kh20N80T3 Alloy after Different Surface Treatments

Surface state	Diffusion coefficient, D, cm²/sec, at various temperatures, °C		
	600°	700°	800°
Electropolished	$0.4 \cdot 10^{-14}$	$0.7 \cdot 10^{-13}$	$0.9 \cdot 10^{-12}$
Annealed	$4 \cdot 10^{-14}$	$6 \cdot 10^{-13}$ (In air)	$4.5 \cdot 10^{-12}$
	—	$0.8 \cdot 10^{-13}$ (In vacuum)	—
Ground	$12 \cdot 10^{-14}$	$13 \cdot 10^{-13}$	$9.2 \cdot 10^{-12}$
Sandblasted	$16 \cdot 10^{-14}$	$15 \cdot 10^{-13}$	$12 \cdot 10^{-12}$

In [10] the author attempted to determine separately the effective coefficient of surface diffusion and the thickness of the layer in which it occurs. He concluded that this layer must consist of about 100 atomic layers. However, we do not know of any investigation in which the diffusional mobility in the thin surface layer was measured directly. Since there are imperfections in the structure at a depth of at least several microns resulting from surface treatment, it seemed necessary to make a special investigation of this problem.

METHOD

The nickel isotope emits soft β-radiation (with an energy of 0.067 MeV). Calculations show that this radiation is absorbed by a layer of iron 7.6 μ thick.

Therefore, the diffusion of nickel even to a very small depth during annealing must lead to a considerable change in the activity. This is the idea which led us to use radioactive nickel to investigate surface diffusion.

It should be noted that it is very important to cover the surface with a very thin layer of radioactive nickel, since the total depth of diffusion is very small. In our investigation the thickness of the layer was less than 1 μ.

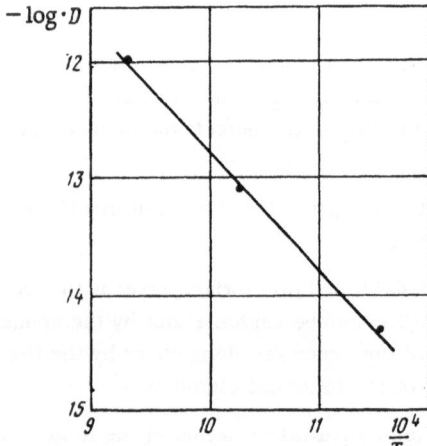

Fig. 1. Temperature dependence of the self-diffusion coefficient of nickel in the Kh20N80T3 alloy after grinding.

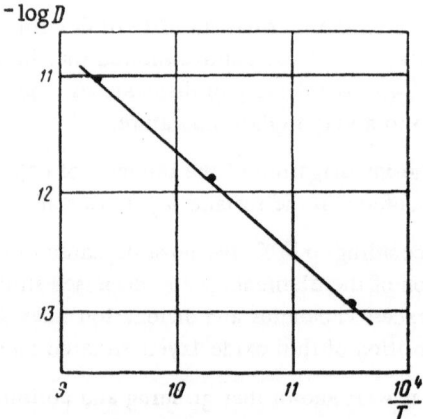

Fig. 2. Temperature dependence of the self-diffusion coefficient of nickel in the Kh20N80T3 alloy after electropolishing.

44

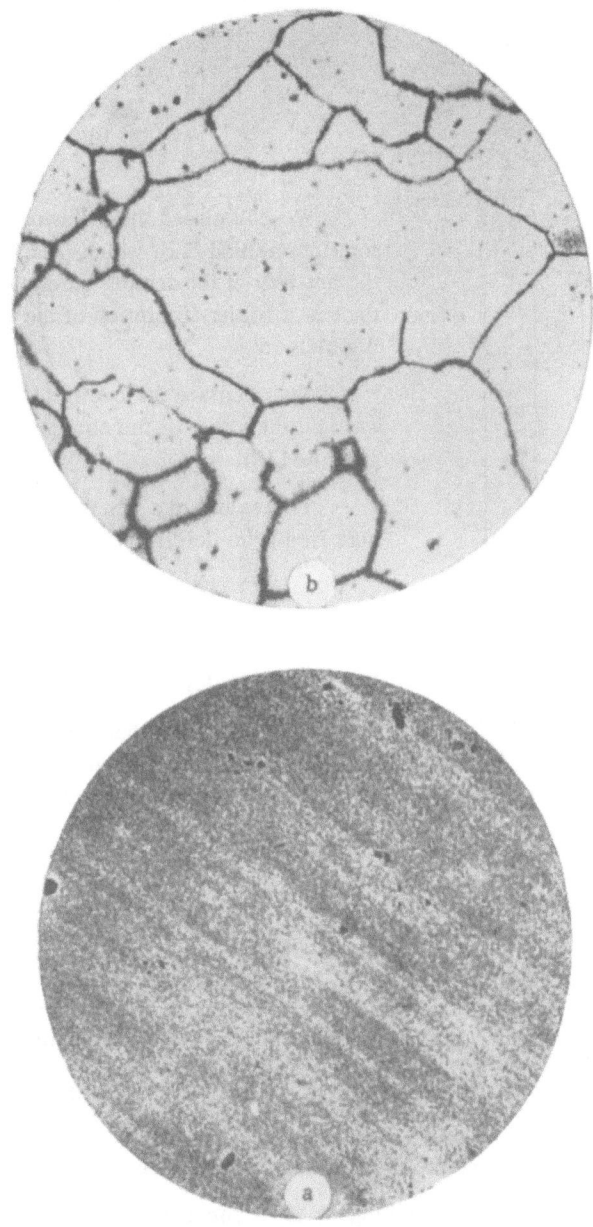

Fig. 3. Distribution of the diffusing element in the surface layer at a depth of 8 μ after different treatments.
a) After grinding; b) after electropolishing. Autoradiograms. ×50.

TABLE 2

Surface state	Q, cal/g–atom
Electropolished	47,000
Annealed	44,600
Ground	39,900
Sandblasted	38,100

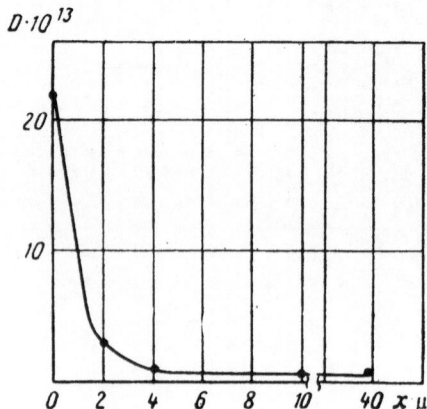

Fig. 4. Variation of the self-diffusion coefficient of nickel with the distance x from the ground surface of the Kh20N80T3 alloy.

We calculated the diffusion coefficient by the absorption method [11], which is based on the difference in the activity of the sample before and after annealing as the result of the absorption of the radiation of the diffusing element.

To calculate the coefficient of self-diffusion of nickel we used the value of the absorption coefficient $\mu = 10^3 \, \text{cm}^{-1}$ given in [12].

The values of μ for the Kh20N80T3 alloy and for pure iron were determined from the relationship

$$\frac{\mu_1}{\rho_1} = \frac{\mu_2}{\rho_2},$$

where ρ is the density.

It was assumed that the coefficient of absorption of the monochromatic radiation of the diffusing element is directly proportional to the density of the sample.

The value of μ is 922 cm^{-1} for the Kh20N80T3 alloy and 883 cm^{-1} for pure iron.

RESULTS

Effect of Surface State on the Self-Diffusion of Nickel in the Kh20N80T3 Alloy

We studied diffusion in the surface layer of the Kh20N80T3 alloy in four different states.

The first state was the initial state, i.e., after electrolytic polishing (a layer of about 30 μ was removed).

The second state was the state resulting from annealing. The samples were annealed 2 h at 900°C in vacuum and in air (the surface of the sample being oxidized in this latter case).

The third state was that resulting from grinding. A layer about 0.1 mm thick was removed from the surface by grinding.

The fourth state was that resulting from sandblasting. The samples were sandblasted 5 min under a pressure of 4-5 atm. Before sandblasting, the surface was polished electrolytically to remove the surface layer in order to bring the surface to the same initial state.

The diffusion coefficients were measured at 600, 700, and 800°C, and in some cases at 500°C.

The Kh20N80T3 alloy recrystallizes at about 800°C, and therefore we chose a temperature of diffusional annealing below the recrystallization temperature but close to the temperature at which this alloy is usually used and also a temperature which would ensure considerable diffusion.

The samples were annealed for 20-25 h at 700-800°C and for 30-35 h at 600°C.

The results are given in Table 1.

TABLE 3. Heterodiffusion Coefficients of Nickel in Iron after Different Surface Treatments

Surface state	Diffusion coefficient, D, cm^2/sec, at various temperatures, °C		
	600°	700°	800°
Electropolished	$1.5 \cdot 10^{-12}$	$4.2 \cdot 10^{-12}$	$1.3 \cdot 10^{-11}$
Annealed	$2 \cdot 10^{-12}$	$6.5 \cdot 10^{-12}$	$1.3 \cdot 10^{-11}$
Ground	$2.4 \cdot 10^{-12}$	$5.6 \cdot 10^{-12}$	$1.3 \cdot 10^{-11}$

These results show that abrasion considerably increases the diffusion coefficient in the surface layer. Thus, at 600°C the diffusion coefficient is about 30-40 times higher, at 700°C it is about 20 times higher, and at 800°C it is about 10 times higher than in the initial state.

The diffusion coefficients of annealed samples are higher than those of electropolished samples. They increase by a factor of 10 at 600-700°C, and a factor of 20 at 800°C. It is characteristic that after annealing in vacuum (30 min and 2 h at 900°C) the diffusion coefficients are the same as after electropolishing.

At 500°C, at which ordinary methods do not reveal any diffusional mobility (self-diffusion in substitution solid solutions), the diffusion coefficient in the surface layer of ground samples was found to be equal to $6.6 \cdot 10^{-14}$ cm^2/sec.

Figures 1 and 2 show the temperature dependence of the self-diffusion coefficient of nickel in the ground and electropolished Kh20N80T3 alloy. The experimental points follow the $\log D = f(1/T)$ line very satisfactorily.

We determined the activation energy of the self-diffusion of nickel from this temperature dependence. The results are given in Table 2.

It should be noted that the activation energy of self-diffusion after electrolyte polishing is lower than the activation energy of the normal volume self-diffusion of nickel, which is equal to 65,000 cal/g-atom. In this case the diffusional mobility was studied in a surface layer 10 μ thick.

The autoradiograms in Figs. 3a and 3b illustrate the self-diffusion of nickel at 800°C at a depth of 8 μ in electropolished (3b) and ground samples (3a). The fundamental difference in the character of the diffusional flow is very clear. In the first case diffusion is almost completely along the grain boundaries and in the second case it is uniform through the whole volume.

We also studied the variation of the diffusional mobility with the thickness of the layer removed from the surface (Fig. 4). The figure shows a sharp decrease of the diffusion coefficient with increasing thickness of the surface layer removed (the layer was removed by electropolishing). After removal of a layer of 2 μ thick, the diffusion coefficient drops by a factor of seven as compared to that in an untreated sample and drops by a factor of 30 when a layer 4 μ thick is removed, but with further polishing, the value of the coefficient remains unchanged.

TABLE 4. Activation Energy of the Diffusion of Nickel in Iron after Different Surface Treatments

Surface state	Q, cal/g-atom
Electropolished	19,600
Annealed	18,300
Ground	16,450

TABLE 5. Self-Diffusion Coefficients of Nickel in Pure Nickel after Different Surface Treatments

Surface state	Diffusion coefficient, D, cm^2/sec, at various temperatures, °C		
	600°	700°	800°
Electropolished	$3.7 \cdot 10^{-15}$	–	$7 \cdot 10^{-13}$
Annealed	$3.2 \cdot 10^{-14}$	$3.5 \cdot 10^{-13}$	$6.5 \cdot 10^{-12}$
Ground	$8.2 \cdot 10^{-14}$	$5.2 \cdot 10^{-13}$	$6.1 \cdot 10^{-12}$
Sandblasted	$2 \cdot 10^{-13}$	$1.4 \cdot 10^{-12}$	$7.3 \cdot 10^{-12}$

The Effect of the Surface State on the Diffusion of Nickel in Iron

To investigate the effect of the surface state on heterodiffusion, we studied the diffusion of nickel in iron at 600, 700, and 800°C, using electropolished, annealed, and ground samples.

The iron was annealed 2 h at 900°C in vacuum. The results are given in Table 3.

The results show that the diffusion coefficients are practically the same in the three cases.

The calculated values of the activation energy of diffusion are given in Table 4.

It should be emphasized that the value of the activation energy is very low (less than 20,000 cal/g-atom, which is lower by a factor of 2.5 than the activation energy Q of self-diffusion of nickel in the Kh20N80T3 alloy).

Effect of the Surface State on the Self-Diffusion of Nickel in Pure Nickel

To explain the fact that the coefficient of the heterodiffusion of nickel in iron is independent of the surface state, we made the following assumptions.

1. At the temperature investigated the recrystallization processes have time to be completed, which explains the lack of effect during treatment.

2. The surface state essentially affects self-diffusion and does not affect diffusion processes.

3. During diffusional annealing even in vacuum the surface of the sample is damaged to the same degree regardless of the preliminary treatment.

4. A phase transformation of iron occurs in the surface layer during the diffusion process because of the high concentration of nickel.

To check our first assumption, we investigated the self-diffusion of nickel in pure nickel, which recrystallizes at about the same temperature as iron. The surface treatments were the same as in the preceding experiments. The results are given in Table 5.

Table 5 shows that grinding and sandblasting have considerable effects on diffusion in the surface layer.

TABLE 6. Activation Energy of the Self-Diffusion of Nickel after Different Surface Treatments

Surface state	Q, cal/g-atom
Electropolished	45,300
Ground	39,100
Sandblasted	32,500

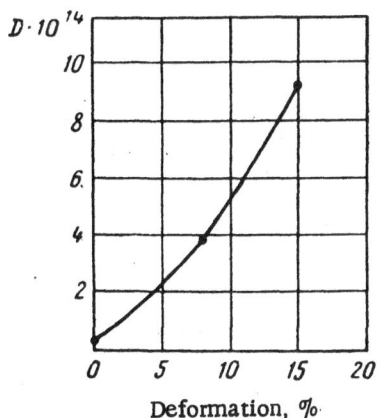

Fig. 5. Variation of the self-diffusion coefficient of nickel at 600°C with the degree of deformation.

The values of the activation energy are given in Table 6.

The table shows that the activation energy Q of self-diffusion of nickel in ground and sandblasted samples is very low and approaches the value of the activation energy of self-diffusion of nickel along the grain boundaries.

DISCUSSION OF RESULTS

Our experiments showed that surface treatment considerably increases the diffusional mobility of the metal. This refers to the self-diffusion of nickel in the complex Kh20N80T3 alloy and also the self-diffusion of nickel in pure nickel. The effect is strongest (D increasing several tens of times) in a thin surface layer—a few microns thick. We can assume that these effects are even greater at the surface itself. Self-diffusion of cobalt in the surface layer of pure cobalt at 1000°C after electrolytic polishing, chemical etching, and grinding was studied in [13]. It was found that the rate of self-diffusion doubles after chemical etching and becomes seven times greater after grinding. The authors explain this effect in terms of the increased scattering of radiation from the irregular surface. However, we think that this phenomenon is of no great importance. Our assumption is confirmed by the following experimental data. We investigated the self-diffusion of nickel in the surface layer at 600°C after 8 and 15% deformation under hydrostatic pressure. This treatment produces no irregularity in the surface, as results from grinding, and yet the coefficient of self-diffusion is greater after this treatment than after electrolytic polishing (Fig. 5).

The diffusional mobility of nickel in iron is not affected by surface cold hardening, as has been shown by studies of the diffusion rate of nickel in iron subjected to grinding (producing an irregular surface) and to surface deformation by hydrostatic pressure (producing no irregularity in the surface). Therefore, we assume that the increase in the diffusional mobility in the surface layers observed in our experiments is due essentially to irreversible changes in the structure and composition in the thin surface layer after grinding, sandblasting, deformation, and annealing in air. This conclusion is confirmed by the autoradiograms (see Fig. 3). In spite of the high-temperature annealing and relatively short annealing time (25-50 h) (conditions in which the Kh20N80T3 alloy is partially recrystallized and pure nickel is completely recrystallized), the effect of surface treatment is still not eliminated. This indicates that the defective surface structure is very stable. This conclusion is of great practical importance in determining the working life of heat-resistant alloys.

It should be noted that annealing in air only partially restores the equilibrium state of the surface of the metal, apparently due to the fact that oxidation of thin layers changes the chemical composition of the metal, and this changes the diffusional mobility. Let us also note that in terms of diffusional mobility sandblasting deforms the surface to the same degree as grinding.

It should also be noted that even after electrolytic polishing the activation energy in the surface layer is lower than the activation energy of ordinary volume self-diffusion (47,000 and 65,000 cal/g-atom, respectively). Possibly this is due to the fact that the removal of a layer several tens of microns thick by electropolishing leads to a surface state which only partly resembles that existing within the crystal lattice.

An important result was obtained in the case of heterodiffusion of nickel in iron. The fact that the surface treatment did not accelerate diffusion in the surface layer cannot be explained by recrystallization processes, since in the case of pure nickel the diffusion is highly accelerated as the result of surface treatment.

The activation energy of the diffusion of nickel in iron is very low (of the order of 16,000-19,000 cal/g-atom) and is close to the value of the activation energy of surface diffusion determined in extremely thin layers (of the order of 10^{-6} cm).

Thus, we have shown by direct experiments that the state of a very thin surface layer differs from that of the bulk metal and that the state of the surface layer depends on the type of surface treatment.

LITERATURE CITED

1. D. M. Evans, Proc. Roy. Soc. 205A (17), 1951.
2. I. A. Gindin and V. S. Kogan, State of the surface layer of zinc single crystals after grinding and annealing, Fiz. Metal. i Metalloved. 5 (2), 1957.
3. Ya. E. Geruzin and A. A. Shpunt, Deformation and destruction of the presurface layer in lithium fluoride crystals, Dokl. Akad. Nauk SSSR 130 (4), 1960.
4. J. J. Gilman and W. G. Johnston, J. Appl. Phys. 27 (9), 1956.
5. A. V. Stepanov, Problems in structural analyses of deformed crystals in view of new data on the mechanism of slip formation, Izv. Akad. Nauk SSSR, Ser. Fiz. 17 (3), 1953.
6. J. J. Gilman and W. G. Johnston, J. Appl. Phys. 30 (2), 1959.
7. I. A. Gindin and V. S. Kogan, State of the surface layer of zinc single crystals after grinding and annealing, Fiz. Metal. i Metalloved. 5 (2), 1957.
8. Ya. E. Geguzin, Dislocational structure of presurface layers of single crystals after grinding, Izv. Vysshei Shkoly, Ser. Fiz., No. 6, 1962.
9. Ya. E. Geguzin and N. N. Ovcharenko, Self-heating of defects on the surface of polycrystalline copper, Fiz. Metal. i Metalloved. 9 (4), 1960.
10. Ya. E. Geguzin, Microscopic Defects in Metals, Metallurgizdat, 1962.
11. S. D. Gertsriken and I. Ya. Dekhtyar, Diffusion in Solid Metals and Alloys, Part III, Fizmatgiz, 1960, p. 134.
12. H. Burgers and R. Smoluchowski, J. Appl. Phys. 26:491, 1955.
13. R. Ruder and C. Birchenall, J. Metals 191 (2): 142-146, 1951.

PART II

RECRYSTALLIZATION AND STRUCTURAL DEFECTS

INVESTIGATION OF THE STATE OF THE GRAIN BOUNDARIES
DURING RECRYSTALLIZATION OF IRON AND IRON ALLOYS

S. Z. Bokshtein, S. T. Kishkin, and L. M. Moroz

Recrystallization is a very general process which can be induced in any given part of an object.

However, the recrystallization mechanism is not yet completely clear. In particular, the mechanism of the rebuilding of grain boundaries during the formation and subsequent growth of new grains resulting from recrystallization is not clear. There is no general concept of the mechanism by which impurities affect the recrystallization process.

Up until recently, recrystallization was studied essentially by metallographic or x-ray analysis and local processes were not completely investigated.

We investigated recrystallization by using radioactive isotopes. We developed a method in which by using autoradiograms we could observe local displacements of atoms on the grain boundaries of the initial grains during plastic deformation and subsequent recrystallization annealing. We made a systematic study of the state of the boundaries during recrystallization in pure iron and iron alloys.

METHOD

We investigated the behavior of iron atoms and impurity atoms on the grain boundaries during recrystallization. To obtain dependable answers to our questions it was necessary to make parallel observations of the same grain metallographically and autoradiographically. The object of the investigation was Sulinsk iron.

Both the autoradiographic and metallographic methods of investigation were used. The samples were coated with a radioactive isotope. As the result of diffusional annealing the radioactive isotope of the base metal or the impurity followed the grain boundaries of the metal investigated. Therefore, it was possible to observe local displacements of atoms on the boundaries during deformation and recrystallization annealing of the samples.

The behavior of the boundaries can be observed only in the very limited layer of the metal in which the radioactive isotope has diffused.

To study the behavior of the atoms of the base metal, the iron samples were coated electrolytically with Fe^{59}. To investigate the behavior of the impurity atoms, the samples were coated with the respective radioactive isotopes of the impurities.

The samples were bars with a section 14×14 mm. They were annealed 9 h at 1250°C to obtain a more uniform structure and large grains, which make the diffusion of the radioactive isotope easier to see.

Samples $10 \times 10 \times 20$ mm were cut from the bars. The samples were polished in a Jaquet electrolyte* to remove the outer cold-hardened layer resulting from mechanical treatment of the bars. The thickness of the cold-hardened layer to be removed was determined from x-rays and microstructural analysis (70-80 μ thick).

---

* The composition of the electrolyte is described in the first article of this collection.

TABLE 1

Initial hardness, HRB	Diffusional annealing conditions	Deformation, %	Hardness after deformation, HRB	Recrystallization conditions	Hardness after re-crystallization, HRB	Repeated deformation, %	Hardness after deformation, HRB	Recrystallization conditions	Hardness after re-crystallization, HRB	Additional annealing temperature, °C	Hardness after heating, HRB
60	800°–90 h	16	78	750°–6 h	60	–	–	–	–	–	–
60	800°–90 h	70	96	750°–4 h	61	–	–	–	–	–	–
61	720°–122 h	14	75	700°–6 h	61	11	78	700°–6 h	61	950°–1 h	60
61	720°–122 h	15	78	1370°–3 h	59	–	–	–	–	–	–
61	720°–122 h	46	95	700°–1 h	61	–	–	–	–	700°–30 h	60
60	720°–122 h	16	80	1200°–1 h	60	–	–	–	–	–	–

An ethyl alcohol solution of picric acid was used to put the microstructure in evidence (4 g of picric acid in 96 ml of ethyl alcohol).

After removal of the cold-hardened layer, a layer of radioactive iron was deposited electrolytically on the polished samples.

The time of electrolytic deposition was 10-15 min, which produced a layer of radioactive iron up to 1.0 μ thick; the activity was 30,000-40,000 pulses/cm$^2 \cdot$min (the activity was measured with the β-counter of the apparatus).

The samples were subjected to diffusional annealing in a vacuum furnace (residual pressure 10^{-3}-10^{-4} mm Hg).

It is important to choose the proper temperature of diffusional annealing to ensure that the radioactive iron diffuses along the grain boundaries.

In our previous work we showed that self-diffusion of iron occurs along the grain boundaries up to very high temperatures (1200°C). However, the higher the temperature the greater the role of volume diffusion.

After a number of preliminary experiments, 720°C was chosen as the temperature of diffusional annealing.

At this temperature the role of boundary diffusion in the total diffusional flow is very great. According to [1], the ratio between the total depth of the self-diffusion of iron and the depth of volume diffusion at 800°C is 1.5-2, whereas at 720°C this ratio is 7-8.

After diffusional annealing the samples were compressed in a 2-ton Amsler press. The relative degree of deformation was determined by the change in the thickness of the sample.

The samples were marked with a diamond stylus to facilitate observation of a given area during the recrystallization process.

The sample was then coated to protect it against corrosion during the photographic exposure. The thickness of this coating did not exceed 1 μ. The sample was then placed on a photographic plate or NIKFI film of the MR type in a desiccator at low temperature for several days. The photographic plates were developed in amidol developer.

The autoradiograms obtained in this way were compared with the micrographs made with an MIM-8 microscope with a magnification of 50 and 100.

In some cases we used the x-ray method. We studied the structure of the sample after deformation and after recrystallization annealing at different temperatures and annealing times.

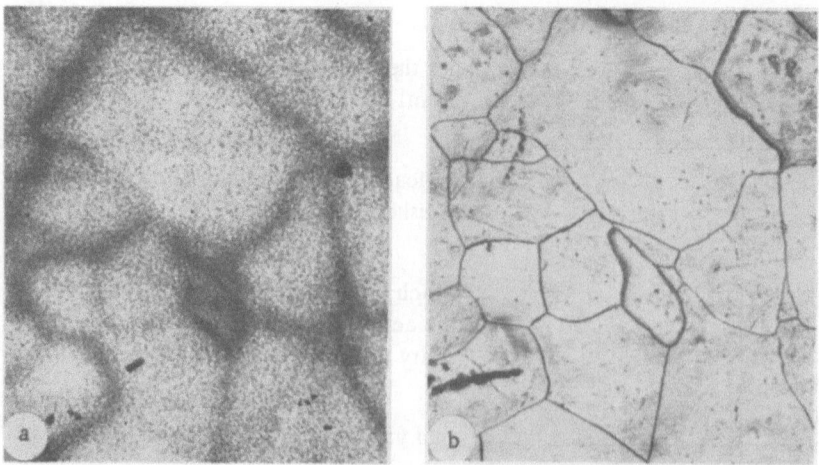

Fig. 1. Structure of the initial iron. ×100. a) Autoradiogram; b) micrograph.

These methods of investigation allowed direct observation of the behavior of grain boundaries during deformation and subsequent recrystallization annealing and of the local displacements of atoms on the grain boundaries.

RESULTS — RECRYSTALLIZATION OF IRON

Let us discuss the results of the investigation of the behavior of iron atoms on the grain boundaries during recrystallization of iron. The iron samples were deformed 10-16 and 45-70%.

The hardness of some of these samples and the conditions of their treatment are given in Table 1.

Recrystallization annealing under the conditions indicated in Table 1 changes the structure considerably.

Autoradiograms and micrographs were made of the same areas of the samples during the different stages of treatment (after deformation and recrystallization annealing). Without exception, the autoradiograms of samples recrystallized after slight (14-16%) and great deformation (46-70%) do not show any change in the microstructure. On the contrary, the micrographs of the same samples show that slight deformation increases the grain size and great deformation decreases the grain size.

X-ray photographs as well as measurements of hardness showed that recrystallization is complete after recrystallization annealing under the conditions indicated in Table 1.

It is of interest to create conditions in which the atoms move a considerable distance from the boundaries of the initial grains. These conditions were produced in two different ways: by plastic deformation and high-temperature recrystallization annealing.

Figures 1 and 2 show autoradiograms and the micrographs of the same area of the sample before and after deformation. Comparison of these figures shows that deformation increases the grain size, although the shape remains the same, and all the grain boundaries are displaced as the result of deformation. In this case there is complete correspondence between the autoradiograms and the micrographs.

The atoms of radioactive iron on the grain boundaries move a considerable distance as the result of plastic deformation. This movement is clear on the autoradiograms (Figs. 1a and 2a) or micrographs (Figs. 1b and 2b).

Recrystallization annealing of the same sample for 1 h at 700°C after 46% plastic deformation produces great changes in the microstructure (Fig. 3b). New grains considerably smaller than the original grains are formed. However, the autoradiogram (Fig. 3a) does not differ from the autoradiogram of the deformed but not annealed sample (Fig. 2a).

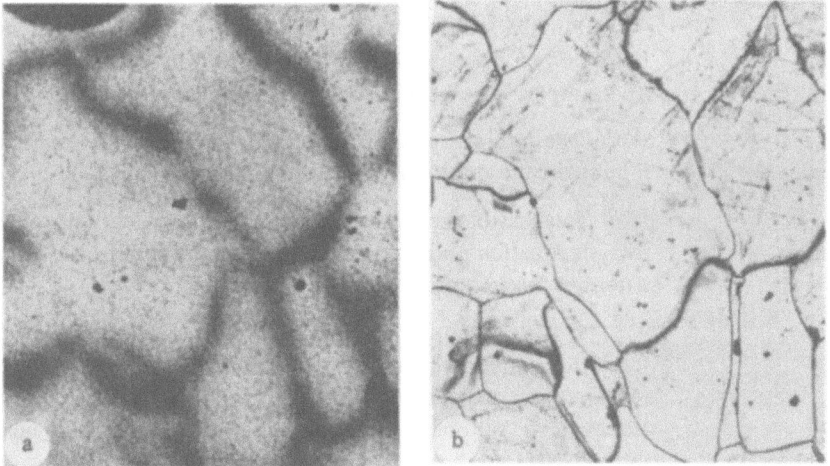

Fig. 2. Structure of iron after 46% deformation. ×100. a) Autoradiogram; b) micrograph.

Additional heating of the same sample at 700°C for 30 min produces changes in the microstructure—the grains grow—although the autoradiogram does not show it.

To study the recrystallization process under the conditions in which the atoms move considerable distances, we investigated the effect of high-temperature annealing. The samples were subjected to 15% deformation and then recrystallization annealing at 1370°C for 3 h. As a result of annealing, the radioactive substance at the grain boundaries becomes distributed almost uniformly through the whole volume of the grain, as can be seen in the autoradiograms.

These data indicate that iron atoms on the boundaries of deformed grains can be displaced relatively great distances by plastic deformation.

We also investigated the behavior of grain boundaries after repeated deformation and subsequent recrystallization.

In this case the autoradiogram shows no change, while the micrograph shows changes after each of these treatments.

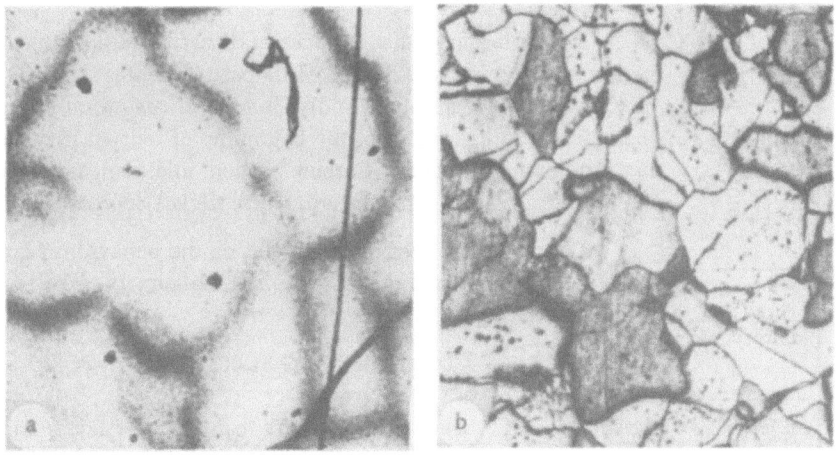

Fig. 3. Structure of iron after 46% deformation and recrystallization at 700°C for 1 h. ×100. a) Autoradiogram; b) micrograph.

TABLE 2

Initial hardness, HRB	Deformation, %	Hardness after deformation, HRB	Recrystallization conditions	Hardness after recrystallization, HRB	Conditions of additional heating				
60	16	80	700°-1 h	60	750°-1 h	950°-45 min	—	—	—
60	52	88	700°-1 h	61	750°-1 h	950°-45 min	—	—	—
60	54	92	700°-1 h	62	750°-1 h	750°-28 h	800°-1 h	950°-45 min	1200°-30 min

Finally, we investigated the effect of the $\alpha \to \gamma$ polymorphic transformation on the behavior of radioactive iron atoms on the grain boundaries. The samples were heated above the A_3 point—950°C—for 1 h.

The micrograph shows that polymorphic transformation changes the microstructure drastically, while the autoradiogram shows no change at all.

Analysis of the data obtained in this way leads us to conclude that the recrystallization process does not induce any significant change in the positions of the iron atoms on the grain boundaries.

It is well known that the structure of steel is changed by plastic deformation and subsequent recrystallization annealing (and even more so by repeated deformation and annealing) and also polymorphic transformation. However, the autoradiogram shows that the atoms on the boundaries of the initial grains are not appreciably displaced in any case.

It should be emphasized that the motion of atoms on the grain boundaries for small distances is independent of the degree of deformation, as shown by the experimental results.

RECRYSTALLIZATION AND THE STATE OF IMPURITY ATOMS

ON IRON GRAIN BOUNDARIES

It is well known that a small amount of an impurity has a great effect on the recrystallization of metals, particularly those which are in the solid solution, although the impurities forming particles of the second phase also have a great effect.

We investigated the change in the positions of impurity atoms on the grain boundaries as the result of recrystallization annealing under the following conditions.

1) When the impurity atoms are practically insoluble in the base metal;

2) when the impurity atoms dissolve in the base metal: and a) increase the recrystallization temperature or; b) decrease this temperature.

It is well known that the recrystallization temperature of iron depends to a great extent on the degree of its purity. It was shown in [2] that the addition of cobalt, chromium, molybdenum, or tungsten to iron considerably increases the temperature of the beginning of recrystallization, the annealing time being the same. Carbon, silicon, and manganese slightly increase the recrystallization temperature, while nickel decreases it.

To study the effect of impurities on the behavior of grain boundaries during recrystallization of iron, we used carbon, tin, tungsten, and nickel.

IMPURITY ATOMS FORMING SUBSTITUTION SOLID SOLUTIONS

Position of Tungsten Atoms on the Grain Boundaries

We chose tungsten because it increases the recrystallization temperature of iron considerably. The addition of 1.5% W increases the activation energy of the recrystallization of iron from 45-55 to 85-90 cal/g-atom.

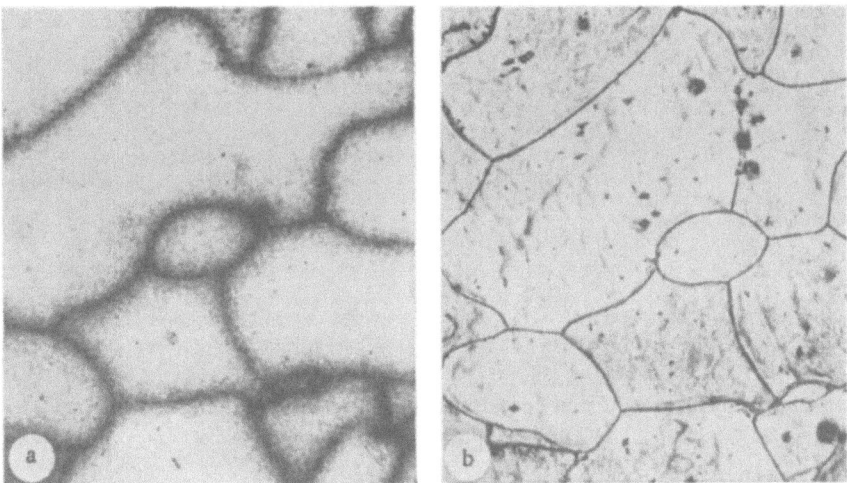

Fig. 4. Structure of iron containing tungsten after 16% deformation. ×100.
a) Autoradiogram; b) micrograph.

The method of investigation used for the study of the recrystallization process of iron containing tungsten was the same as that used for pure iron.

Iron samples were coated electrolytically with radioactive tungsten and then subjected to diffusional annealing in a vacuum furnace at 700°C for 120 h. This temperature ensures preferential diffusion of tungsten along the grain boundaries of iron.

The samples were deformed 10-15 and 50-70%.

The hardness and treatment conditions for three samples are given in Table 2.

The autoradiograms of the samples during different stages of the treatment show without exception that the recrystallization of samples deformed either slightly or greatly produces no change in the microstructure of the metal, while the micrographs of the same areas show drastic changes in the microstructure.

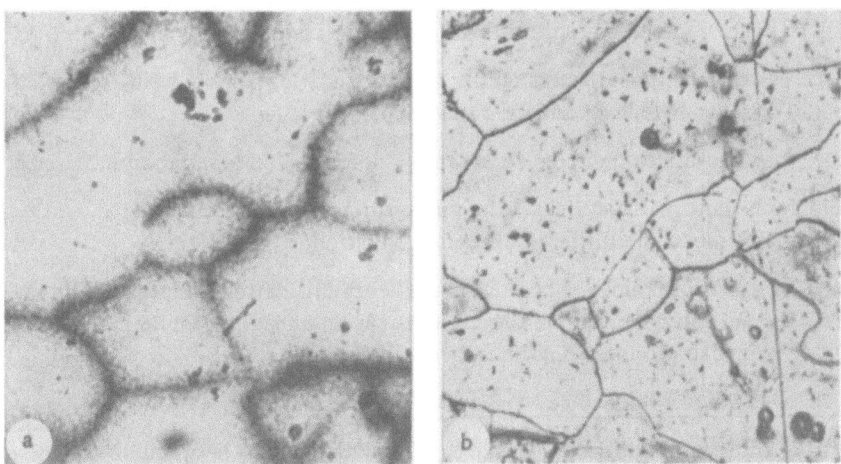

Fig. 5. Structure of iron containing tungsten after 16% deformation and recrystallization for 1 h at 700°C. ×100. a) Autoradiogram; b) micrograph.

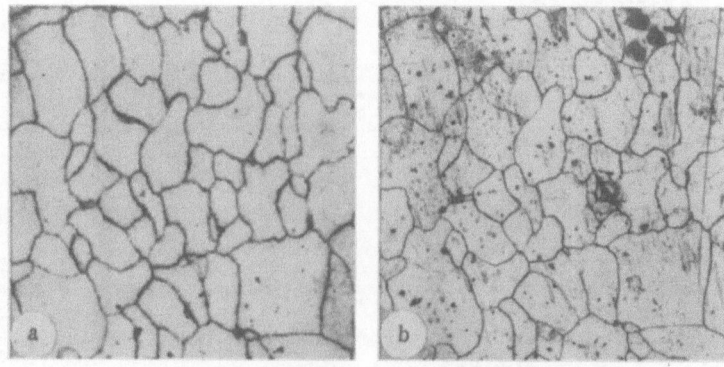

Fig. 6. Structure of iron containing nickel after 56% deformation. ×50.
a) Autoradiogram; b) micrograph.

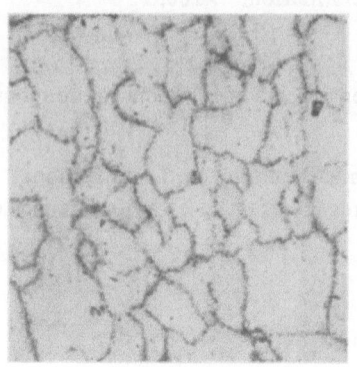

Fig. 7. Autoradiogram of iron containing nickel after 56% deformation and recrystallization at 700°C for 1 h. ×50.

Figures 4 and 5 show the structure of the same area of the sample after 16% deformation and after subsequent recrystallization at 700°C for 1 h.

The autoradiogram shows that after heating of the sample deformed 54% to 1200°C the tungsten atoms remain in practically the same places on the boundaries of the grains of the deformed metal before heating.

Thus, recrystallization and subsequent high-temperature heating do not induce any significant changes in the position of the tungsten atoms on the grain boundaries of iron.

The behavior of nickel is similar.

Position of Nickel Atoms on the Grain Boundaries

It is well known that nickel decreases the recrystallization temperature of iron. The solubility of nickel in iron is infinite. Iron was coated electrolytically with a layer of radioactive nickel and then subjected to diffusional annealing at 700°C.

We studied the behavior of nickel atoms on the grain boundaries of iron during recrystallization after slight deformation (14-18%) and great deformation (50-56%).

Figures 6 and 7 show the microstructure of the sample deformed 56% and recrystallized at 700°C for 1 h.

The autoradiograms show that in all cases the microstructure remains unchanged, while the micrographs show changes in the microstructure resulting from recrystallization.

To determine the effect of prolonged heating at the recrystallization temperature on the behavior of nickel atoms in iron, we subjected the sample to recrystallization annealing for 78 h and 1.5 h. After 78 h at 700°C the positions of the nickel atoms had not changed.

Position of Tin Atoms on the Grain Boundaries

The solubility of tin in iron is rather high (the solubility being 10% between 800°C and room temperature). However, the structure of tin atoms is very different from that of iron atoms, and therefore the addition of tin induces great deformation in the crystal lattice of iron.

Iron was coated electrolytically with radioactive tin and then subjected to diffusional annealing at 700°C, a temperature ensuring the penetration of tin along the grain boundaries.

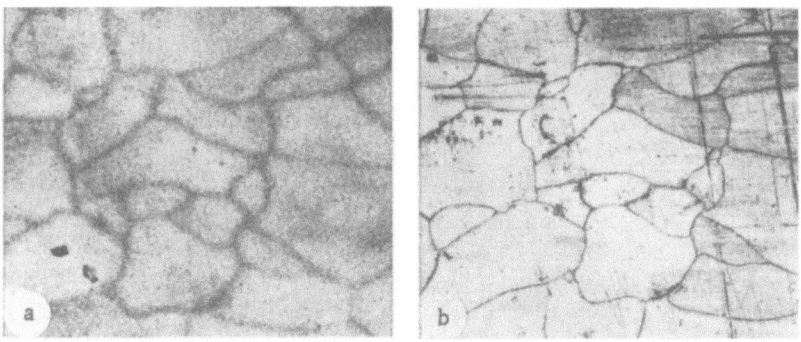

Fig. 8. Structure of iron containing tin after 54% deformation. ×50.
a) Autoradiogram; b) micrograph.

Figures 8 and 9 show the microstructure of the sample after 54% deformation (Fig. 8) and after subsequent recrystallization at 700°C for 1 h (Fig. 9).

The autoradiograms and micrographs show that after both low and high degrees of deformation recrystallization does not induce any change in the positions of tin atoms.

The behavior of tin during recrystallization of iron is similar to that of nickel and tungsten.

IMPURITIES FORMING INTERSTITIAL SOLID SOLUTIONS

We investigated technically pure iron containing radioactive carbon. We chose carbon because it is practically insoluble in α-iron—the solubility of carbon in α-iron at 600°C is 0.01%; it is 0.02% at 700°C [3]. Also, it is well known that carbon forms an interstitial solid solution in α-iron.

According to [3], 0.02-0.03% carbon in iron exceeds the solubility limit of carbon in ferrite, and as a result it prevents collecting recrystallization and grain growth, particularly between temperatures of 620 and 700°C.

The samples were saturated with carbon from donors at 700°C for 2 h. This type of saturation ensured preferential distribution of carbon on the grain boundaries. The donors were iron plates 10 × 20 × 2 mm. The donors were saturated with radioactive carbon from barium carbonate in sealed quartz ampules at 970°C.

We studied the recrystallization of samples deformed 10-15 and 50-70%.

The conditions of treatment and the hardness of two samples are given in Table 3.

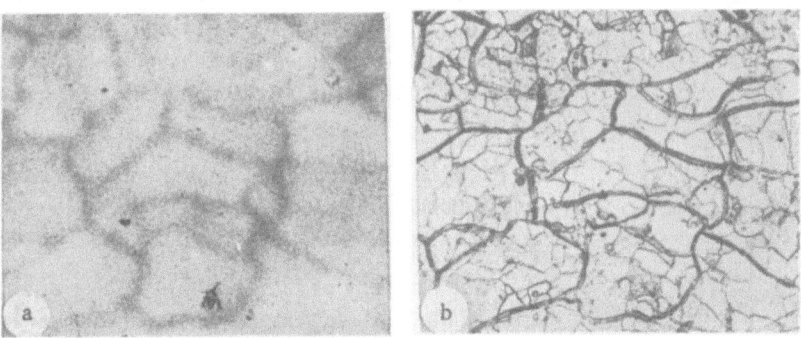

Fig. 9. Structure of iron containing tin after 54% deformation and recrystallization
at 700°C for 1 h. ×50. a) Autoradiogram; b) micrograph.

TABLE 3

Initial hardness, HRB	Deformation, %	Hardness after deformation, HRB	Heating temperature, °C	Hardness after heating, HRB	Recrystallization conditions	Hardness after recrystallization, HRB	Conditions of additional heating	
61	13	78	650°–45 min	66	700°–45 min	62	750°–30 min	950°–20 min
62	54	93	550°–30 min	91	650°–45 min	62	750°–30 min	950°–20 min

Recrystallization does not occur at temperatures below 550°C, as can be seen by the fact that the hardness after deformation (HRB 93) changes only to HRB 91 after annealing at 550°C. Apparently, however, the grains of the α-phase do grow at this temperature.

Further heating to 650°C induces practically complete recrystallization, which can be seen in the changes of the hardness and the microstructure (Figs. 10b and 11b).

The carbon remains along the grain boundaries of unrecrystallized grains formed after heating at 550°C.

Further heating at 750°C leads to the growth of recrystallized grains (Fig. 12b) and to displacement of carbon atoms to the boundaries of the new grains (Fig. 12a). However, carbon does not move to the boundaries of all the new grains. In many cases the carbon network borders a few small grains. This results either from the lack of carbon or the fact that the recrystallization process is not completed during this period of time.

Possibly the different structures of the boundaries also have an effect—carbon atoms first occupy the boundaries of the grains with the least perfect structure.

Recrystallization annealing of samples deformed only slightly produces insignificant changes in the microstructure. Further heating (750°C for 30 min) increases the effect: separate grains grow, some grains decrease in size, some grains change their shapes completely. The autoradiogram shows these changes.

The autoradiograms and micrographs show that recrystallization of slightly deformed samples induces the displacement of carbon atoms from the boundaries of deformed grains to the new boundaries of the recrystallized grains.

We also studied the effect of the polymorphic $\alpha \rightarrow \gamma$ transformation on the behavior of radioactive carbon atoms on the grain boundaries. For this purpose the samples were heated to a temperature above the A_3 point (950°C) for 1 h.

The autoradiogram and micrograph (Fig. 13) show the changes in the microstructure resulting from the polymorphic transformation. Completely new grains are clearly visible. The autoradiogram (Fig. 13a) shows that the carbon atoms have moved to the boundaries of the new grains and formed large accumulations of carbide—the separate dark spots.

It is characteristic that the carbon atoms always move toward the grain boundaries rather than in the direction of the decrease of the concentration of the impurity.

The tendency of carbon to follow the grain boundaries can also be seen in the case of samples subjected to a low degree of deformation, for which the distribution of the carbon atoms on the autoradiogram always corresponds to the position of the boundaries on the micrographs.

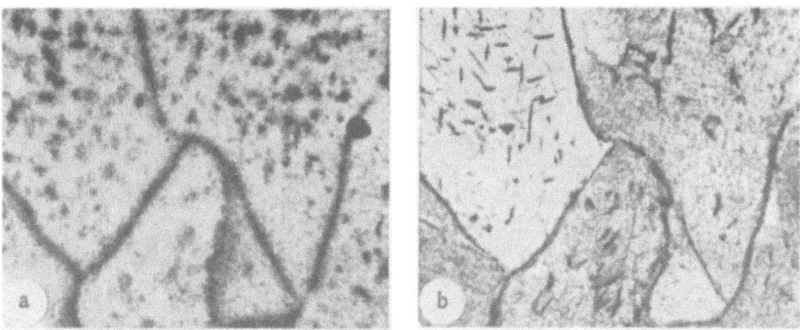

Fig. 10. Structure of iron containing carbon after 54% deformation. ×100.
a) Autoradiogram; b) micrograph.

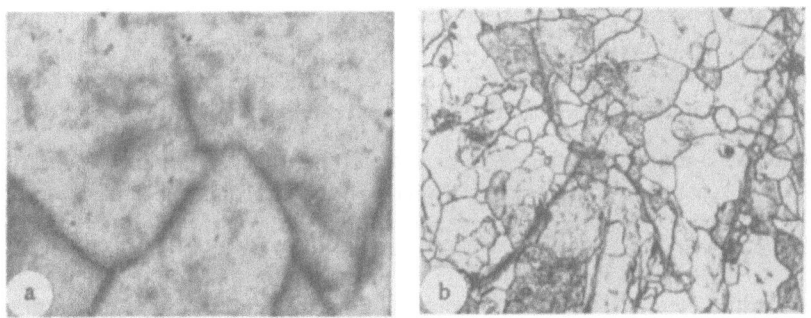

Fig. 11. Structure of iron containing carbon after 54% deformation and recrystallization
at 650°C for 45 min. ×100. a) Autoradiogram; b) micrograph.

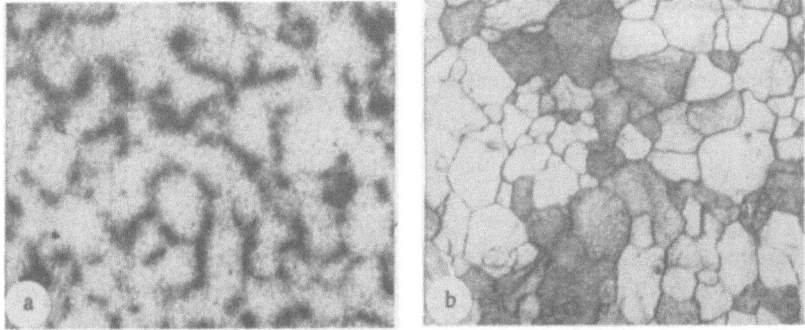

Fig. 12. Structure of iron containing carbon after 54% deformation and repeated recrystallization at 750°C for 30 min. ×100. a) Autoradiogram; b) micrograph.

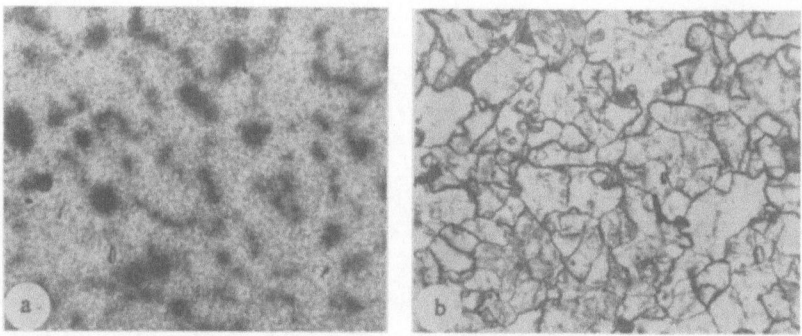

Fig. 13. Structure of iron containing carbon heated at 950°C for 1 h. ×100.
a) Autoradiogram; b) micrograph.

DISCUSSION OF RESULTS

This investigation showed that the recrystallization process produces no significant changes in the positions of the atoms on the grain boundaries.

At the same time, the structure of steel subjected to plastic deformation and subsequent recrystallization annealing, polymorphic transformation, and (especially) to repeated deformation and annealing changes quite drastically. Completely new grains with new boundaries occur. The grains become larger or smaller than the original grains, depending on the degree of deformation and the recrystallization conditions.

The recrystallization process and the grain growth during heating occur under conditions in which the atoms at the surfaces of separation of crystals are displaced very small distances, and consequently the atoms of the boundaries of deformed grains are not the structural material from which the boundaries of new recrystallized grains are formed.

At the present time, there are no general theories of the mechanism of grain growth or the rebuilding of grain boundaries during recrystallization.

According to the views of a number of investigators, grain growth is related essentially to the displacement of single atoms from deformed grains to new, growing grains [4].

The concept of the diffusional nature of grain growth is based on the fact that the activation energies of recrystallization and diffusion or self-diffusion are very similar [5].

The value of the activation energy of recrystallization has no clear physical meaning, since this complex process includes a number of elementary processes. It must also be emphasized that the experimental value of the constant A_G in the equation for the growth rate G

$$G = A_G e^{\frac{-Q_G}{RT}}$$

(where Q_G is the activation energy of grain growth) is 10^{10} times greater than the theoretical value calculated on the assumption that recrystallization occurs by the ordinary diffusion mechanism. In [6, 7] it was concluded that grain growth cannot be based on the diffusion mechanism and it was assumed that several atoms are activated simultaneously in the elementary process.

Thus, Lücke [7] assumed that areas of the crystal lattice containing $\sim 10^{10}$ atoms adhere to the lattice of the growing crystal at a high rate.

Schaefer, who studied the recrystallization of copper, and other investigators came to a similar conclusion.

Probably the displacement of boundaries during recrystallization and subsequent grain growth is related to some specific mechanism which differs from the ordinary diffusion mechanism.

Possibly shear stresses occur in the deformed crystal during recrystallization and as the result of these stresses atomic groups are collectively displaced small distances (of the order of interatomic distances), while the boundaries are displaced considerable distances at a high rate.

This picture corresponds to Burgers' model [8], who assumed that the motion of boundaries and grain growth during heating are the result of the displacement of dislocations.

The role of impurities in recrystallization must be discussed separately.

It is well known that impurities and alloyed elements have a considerable effect on the kinetics of the processes occurring during annealing of plastically deformed metals and alloys. Both insoluble and soluble impurities affect the recrystallization process and grain growth, the effect of soluble impurities being much greater.

Most impurities decelerate recrystallization, but to different degrees. In this respect one must distinguish the effect of impurities on two types of elementary processes—the formation of nuclei and the growth of nuclei.

Many attempts at a theoretical interpretation of the effect of soluble impurities on the growth of re-crystallization centers have been made. These attempts can be separated into two basic types.

Investigators favoring the first type base their explanation on the assumption that the impurities are dis-solved uniformly in deformed metal and assume that the impurity affects the whole volume of the metal. The others assume that impurities have local effects.

Bochvar, from an investigation of the recrystallization of tin alloys, concluded that there is a definite relationship between the effect of the impurities on the growth rate of recrystallization centers and the solu-bility of impurities in the solid phase.

In [9] it was indicated that there is a correlation between the effect of alloying elements on the recrys-tallization temperature and the atomic radius and valence of the impurity.

There is also a point of view that impurities which increase the binding energy of the atoms of the base metal, and consequently decrease the general level of mobility of atoms, slow down recrystallization. How-ever, according to [10], this factor cannot be decisive.

It should be emphasized that although most impurities have little effect on the mobility of the atoms of the base metal they do slow down recrystallization to a certain extent. Also, even the alloyed elements which do have a great effect can decrease the general mobility of atoms only by one or two orders within the range of recrystallization temperatures, while the growth rate of recrystallization centers sometimes changes by several orders.

This is particularly true for elements which, according to autoradiograms and other data, are strongly horophilic. Therefore, one can assume that this phenomenon is related to the irregular distribution of impuri-ties within the volume of the deformed alloy.

Small amounts of impurities have the greatest effect. The optimum concentration is of the order of 0.01%, which is considerably below the solubility limit. This effect of a small amount of impurities is appar-ently due to their irregular distribution within the volume of the crystal.

Therefore, it was assumed that impurities have local effects. In fact, according to many investigators, impurities are often concentrated along separation surfaces [11].

Lücke [12] attempted to explain the enrichment of wide-angled boundaries in impurities and the de-crease of the mobility of impurities in terms of the Cottrell clouds around the dislocations forming these bound-aries. The same conclusion was reached in [13] with respect to boundaries with sharp angles.

It was shown experimentally in [14] that the increase in the concentration of impurity atoms at the bound-ary of the growing recrystallization center slows down its growth. The author explained this phenomenon by assuming that recrystallization centers grow without any significant displacement of atoms as the result of re-building without any movement.

In [15] it was assumed that the growth of nuclei is slowed down because the impurity atoms are removed by diffusion from the front of the moving boundary. The impurity atoms which are on the grain boundaries in the deformed metal penetrate the grains during growth of the nuclei, and this leads to an increase of the total free energy. As the result, the impurity atoms tend to cross the boundary again. According to [15], the boundaries move at the same time as the impurity atoms are removed by diffusion.

The impurities forming particles of the second phase slow down (mechanically) the motion of the boundaries.

We have shown that the atoms of soluble impurities—nickel, tungsten, tin—like the atoms of the base metal, remain practically in their original positions during the recrystallization of iron in spite of the considerable changes in the microstructure of the metal. Long periods at the recrystallization temperature (78 h at 700°C for iron containing nickel, 30 h at 700°C for iron containing tin, and 28 h at 750°C for iron containing tungsten), heating at temperatures above the A_3 point, and also high-temperature annealing (1200°C for 30 min for iron containing tungsten) do not induce the displacement of the impurity atoms from their original positions to the new boundaries of recrystallized grains.

This result is independent of the degree of deformation.

Our data show that the formation of boundaries of newly recrystallized grains is very rapid after a high degree of deformation and apparently is independent of the motion of the impurity atoms during recrystallization.

The analysis of the results of our investigation showed that the impurity atoms on the grain boundaries which form interstitial solid solutions behave in a very different way from the atoms of the base metal and the impurity atoms forming substitution solid solutions.

Unlike iron, tungsten, nickel, and tin atoms, carbon atoms follow the boundaries of the newly formed grains. It is only the difference in the diffusion rates of carbon on one hand and the recrystallization rate on the other that leads to the fact that the positions of carbon atoms and the positions of the boundaries of recrystallized grains do not coincide at given stages

The new grains recrystallize at a high rate and the carbon atoms do not have time to occupy the boundaries of these new grains during the initial stages of the process. The carbon atoms move to the new boundaries only during further development of the grains.

It is characteristic that carbon atoms always move to the boundaries of the grains and not in the direction in which the concentration of the impurity decreases.

The tendency of carbon to follow the grain boundaries can be seen also when analyzing the effect of small degrees of deformation. When the degree of deformation is low the positions of the carbon atoms and the positions of boundaries coincide.

The fact that the carbon moves toward the new boundaries is apparently due to the fact that carbon is horophilic, while the fact that the carbon atoms finally occupy the boundaries of the recrystallized grains is due to their high mobility.

During the growth of the α-phase, which occurs before recrystallization, the carbon atoms follow the boundaries of the ferrite grains.

It would be interesting to calculate the activation energy of the transfer of carbon atoms to the new boundaries and to compare it with the activation energy of the diffusion of carbon in iron.

LITERATURE CITED

1. C. Lemoine and P. Lacombe, Rev. Met. 54 (9): 653, 1957.
2. L. I. Kogan and R. I. Éntin, Problems of Metal Science and the Physics of Metals, Collection No. 2, Metallurgizdat, 1951, p. 216.
3. B. I. Bruk, Metal Science, Collection of Articles, Sudpromgiz, 1959, p. 326.

4. C. Peterson, Z. Metallk. 38: 289, 1947; W. A. Anderson and R. F. Mehl, Trans. AIM, Inst. Met. Div. 161: 140, 1945; P. L. Gruzin, Metalloved. i Term. Obrabotka, No. 10, 1960.

5. G. W. Wench and H. Walker, Trans. ASM, 44: 1185, 1952; V. Zait, Diffusion in Metals [Russian translation], IL, 1958, p. 280; D. Turnbull, Trans. AIMME, 191: 661, 1951.

6. N. F. Mott, Proc. Phys. Soc. 60: 391, 1948.

7. K. Lücke, Z. Metallk. 41: 114, 1950.

8. W. G. Burgers, Z. Metallk. 41: 2, 1950.

9. K. Lücke and K. Detert, Acta Met. 5 (11): 628, 1957.

10. L. N. Larikov, Vopr. Fiz. Metal. i Metalloved., Akad. Nauk SSSR, 1959, p. 121; 1960, p. 61.

11. V. I. Arkharov, Tr. Inst. Fiz. Metal. AN SSSR, Ural'sk. Filial (8): 54, 1946; S. Z. Bokshtein, S. T. Kishkin, and L. M. Moroz, Investigation of the Structure of Metals by Radioactive Isotopes, Oborongiz, 1959.

12. K. Lücke and K. Detert, Acta Met. 5 (11): 628, 1957.

13. S. S. Gorelik, Recrystallization of Metals and Alloys, Dissertation, MIS, 1962.

14. E. É. Zosimchuk, G. V. Kurdyumov, and L. N. Larikov, Izv. Akad. Nauk SSSR, Ser. Fiz. 23 (5): 615, 1959.

15. S. S. Gorelik and É. P. Spektor, Izv. Vysshikh Uchebn. Zavedenii, Ser. Chernaya Met. (9): 120, 1960.

STUDY OF THE STRUCTURE OF GRAIN BOUNDARIES
DURING RECRYSTALLIZATION OF NICKEL

M. A. Gubareva

We used the method described in the preceding article to investigate the structure of the grain boundaries.

We investigated technically pure nickel which contained 99.7% Ni, 0.1% Co, 0.04% Fe, 0.06% Cu, 0.062% Si, 0.04% C, and 0.005% S. To obtain a uniform structure and to increase the grain size, nickel bars were annealed 1 h at 1100°C. Samples 10 × 10 × 20 mm were then cut from these bars.

The samples were electropolished in a Jaquet electrolyte to remove the cold-hardened outer layer. Then a layer of radioactive nickel up to 1 μ thick was deposited electrolytically. The samples were subjected to diffusional annealing in a vacuum furnace at 800°C. During diffusional annealing radioactive nickel follows the grain boundaries, thus making it possible to observe the local movements of atoms on the grain boundaries during deformation and subsequent recrystallization.

The samples were deformed by compression. The degree of deformation was determined by the change in the thickness of the sample. The hardness was measured before and after deformation and also after recrystallization.

Autoradiograms and the micrographs were made of the same areas of the samples, which were marked beforehand. The micrographs were made with an MIM-8 microscope with a magnification of 50.

The hardness of some of the samples and the conditions of treatment are given in the table.

The autoradiograms of samples after deformation and recrystallization showed no change in samples deformed either slightly (4.3-20%) or greatly (57%). At the same time, the micrographs show considerable change in the microstructure resulting from recrystallization: the grains become larger after low degrees of deformation (4.3%) and become smaller after 20 and 57% deformation.

Figure 1 shows the autoradiogram of a sample after 4.3% deformation, while Fig. 2 shows the autoradiogram of the sample after recrystallization at 900°C for 1 h.

Hardness of Samples and Conditions of Treatment

Initial hardness of the sample, HRB	Deformation, %	Hardness after deformation, HRB	Recrystallization conditions	Hardness after recrystallization, HRB
18	4	43	900°−1 h	18
9	20	67	900°−1 h	7
10	57	92	800°−1 h	19

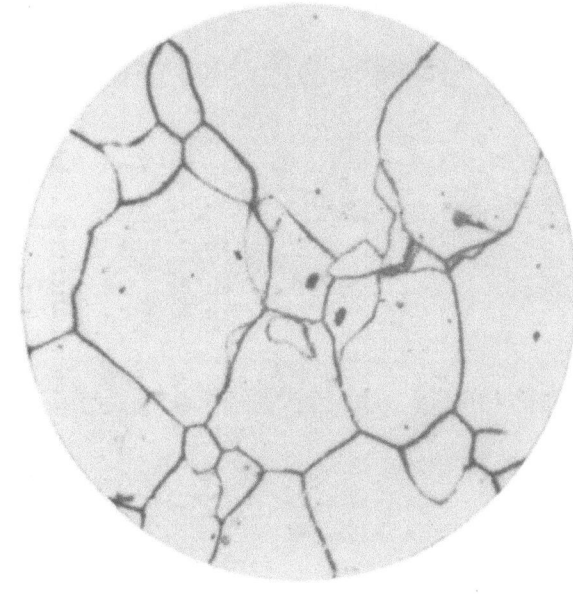

Fig. 2. Structure of nickel after 4.3% deformation and recrystallization annealing for 1 h at 900°C. Auto-radiogram. ×50.

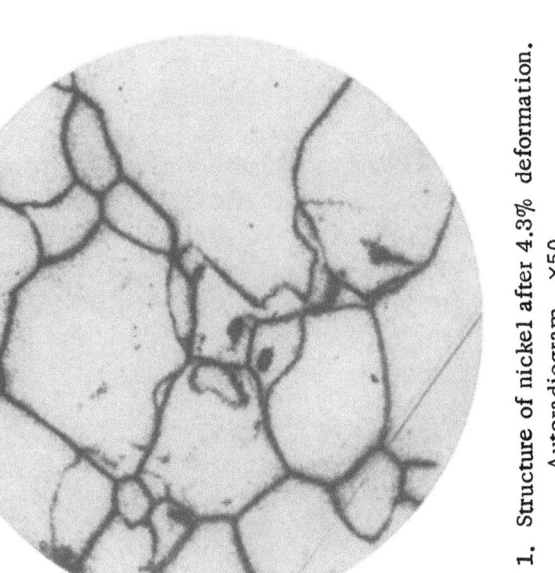

Fig. 1. Structure of nickel after 4.3% deformation. Autoradiogram. ×50.

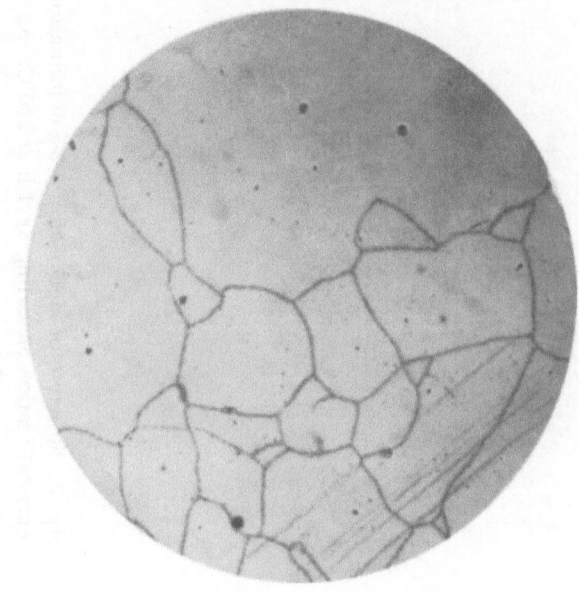

Fig 4. Structure of nickel after 20% deformation and recrystallization annealing for 1 h at 900°C. Autoradiogram. ×50.

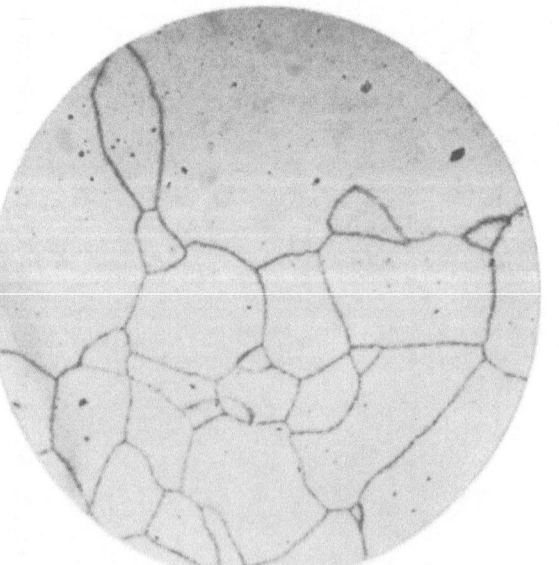

Fig. 3. Structure of nickel after 20% deformation. Autoradiogram. ×50.

The autoradiograms in Figs. 3 and 4 show no changes in the microstructure after 20% deformation and re-crystallization at 900°C for 1 h, while the micrographs show considerable changes in the microstructure.

Deformation of 57% induces movement of atoms to a certain distance from the boundaries of the initial grains. The autoradiograms and micrographs show that although the shape remains the same the grains become larger after deformation, and therefore all the boundaries are displaced. Recrystallization annealing at 800°C for 1 h changes the microstructure considerably. The micrograph shows that the new grains are much smaller than the initial grains, while the autoradiogram shows no change resulting from recrystallization.

Thus, nickel is subject to the same laws as those previously established for pure iron: during recrystallization the atoms on the grain boundaries of the base metal remain in practically the same position as after as much as 60% deformation.

STRUCTURAL IMPERFECTIONS RESULTING
FROM RECRYSTALLIZATION OF METALS

S. Z. Bokshtein, S. T. Kishkin, L. M. Moroz, and V. S. Chaplygina

It is well known that the properties of metals and the processes occurring in metals depend to a large extent on the degree of perfection of the structure. However, the conditions which induce imperfections in the structure and those which remove imperfections are not clear.

The study of the stability of structural and concentrational heterogeneities during recrystallization is of great interest.

It has been shown[*] that the atoms of the base metal (iron, nickel as well as of the impurity atoms (tungsten, nickel, tin) on the boundaries of deformed grains do not move to the boundaries of new grains formed during recrystallization but remain almost in their initial positions; impurity atoms forming interstitial solid solutions (carbon, for example) move from the boundaries of deformed iron grains to the boundaries of new grains formed during recrystallization.

One can assume that the impurity atoms move easily from the grain boundaries into the grains themselves as the result of recrystallization. In fact, since the impurity atoms remain in their initial positions during recrystallization, there should not be any impurity atoms on the new grain boundaries.

It has been possible, in fact, to eliminate the harmful effects of impurities in some cases. Thus, Ovesepyan, who investigated molybdenum—zirconium alloys, showed that the alloys become more ductile after heating at temperatures somewhat above the recrystallization temperature. This improvement occurs only in alloys containing impurities.

Since recrystallization does not always have a positive effect, one can assume that the specific structure of the areas of the crystal corresponding to grain boundaries before recrystallization remains the same after recrystallization, i.e, it must be determined whether there is a "heredity"—a singular "memory"— in a crystal with respect to the initial boundaries and how long this "memory" persists.

We investigated whether the initial grain boundaries remain in the crystal after recrystallization and the growth of new grains. In this investigation we used autoradiographic and micrographic analysis.

The degree of perfection of separate areas of the structure after recrystallization was measured in terms of the depth of diffusion, since a defective structure facilitates diffusion.

The samples were saturated with radioactive elements from special donors after recrystallization.

The stability and the degree of imperfection of the original grain boundaries after recrystallization was studied as a function of the degree of purity of the material and the recrystallization conditions.

We investigated Sulinsk iron and also iron in which the boundaries of the initial grains were enriched in tin and tungsten by diffusion.

The stability of the defective structure during recrystallization was studied in the following way.

---

[*] See Bokshtein, Kishkin, and Moroz, this volume p. 52.

Conditions of Treatment

Material	Initial hardness, HRB	Deformation, %	Hardness after deformation, HRB	Recrystallization conditions	Hardness after recrystallization, HRB	Conditions of diffusional saturation
Iron	60	55	97	650°–45 min	67	600°–1 h
Iron	60	56	97	700°–30 min	66	600°–1 h
Iron	59	54	95	750°–1 h	65	600°–1 h
Iron + tin	60	50	94	650°–45 min	66	600°–1 h
Iron + tin	61	51	94	700°–30 min	67	600°–1 h
Iron + tin	60	51	95	750°–1 h	66	600°–1 h

Iron bars were annealed 9 h at 1250°C and samples 10 × 10 × 20 mm were cut from these bars.

The cold-hardened outer layer was removed by electropolishing in a Jaquet electrolyte. The samples were then chemically etched in a 4% solution of picric acid in ethyl alcohol to reveal the microstructure.

Some of the samples of Sulinsk iron were saturated with tin and tungsten in a vacuum furnace at 700°C before being deformed. Then a layer of a given thickness was removed from the surface by electropolishing, the thickness of the layer corresponding to the depth of volume diffusion, and the samples were then deformed by compression.

The deformed samples were subjected to recrystallization annealing in a vacuum furnace at 650, 700, and 750°C. Metallographic analysis and measurements of the hardness showed that recrystallization occurred at all three temperatures.

Then the samples were saturated with radioactive carbon at 600°C for 1 h. The diffusion temperature was below the recrystallization temperature in all cases, and ensured preferential penetration of carbon atoms along the grain boundaries without any recrystallization.

Then autoradiograms of the surfaces of the samples were made. The degree of imperfection in the structure in separate areas of the sample was determined from the degree of darkening on the autoradiograms. The autoradiograms were compared with the micrographs.

The treatment conditions and the hardness of some of the samples are given in the table.

Our investigation showed that diffusion of carbon is different in iron recrystallized at different temperatures.

In the case of recrystallization at 650°C for 45 min carbon moves preferentially along the initial grain boundaries of the deformed sample, although the micrograph shows only new grain boundaries.

After recrystallization at higher temperatures, i.e., 700°C for 30 min, a double network of boundaries can be seen on the autoradiogram. This double network indicates the penetration of carbon through the areas of the grain where boundaries existed before recrystallization and along the newly formed boundaries.

In the case of recrystallization at 750°C for 1 h the carbon forms a single network along the grain boundaries, but this network consists of the boundaries of the new grains resulting from recrystallization.

The autoradiograms and micrographs illustrate these results.

In Fig. 1, the dark areas on the autoradiogram (1c) (recrystallization at 650°C) correspond to the positions of the old boundaries of deformed grains (1a) and are not related to the boundaries of new grains resulting from recrystallization (1b).

The autoradiogram of a sample recrystallized at 700°C for 30 min is very interesting (Fig. 2c). Here, the double network of boundaries is very clear. Comparison with the micrograph of the same area of the sample after recrystallization (Fig. 2b) shows that the dark borders

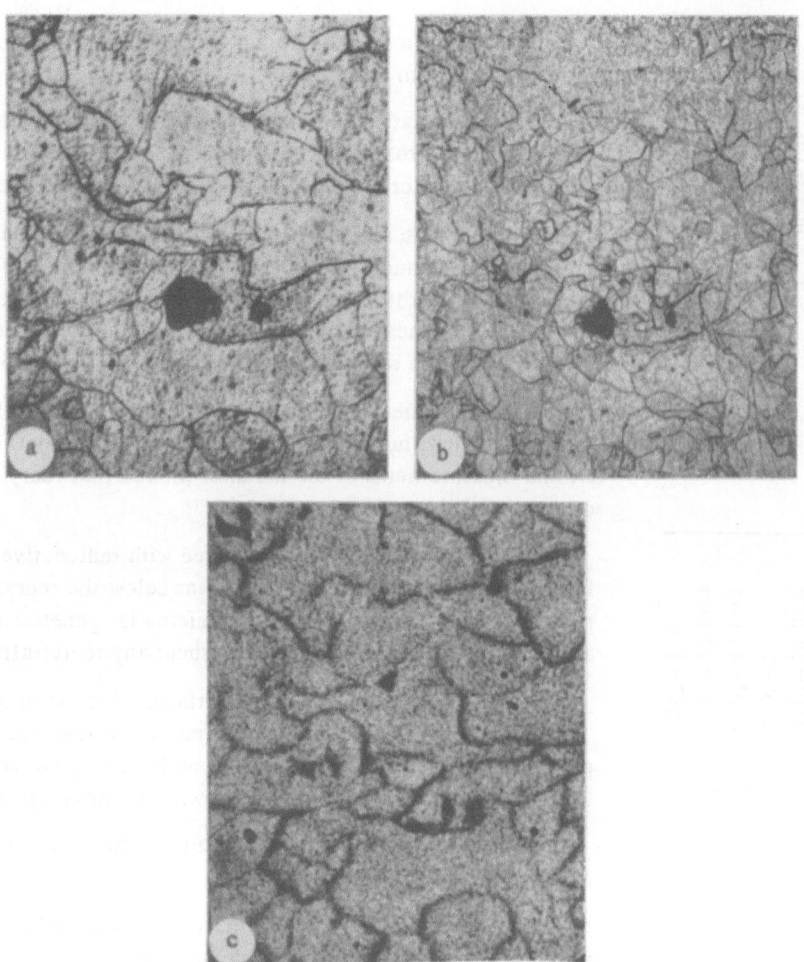

Fig. 1. Structure of iron. ×50. a) Microstructure after 55% deformation; b) microstructure after recrystallization at 650°C for 45 min; c) autoradiogram after recrystallization at 650°C for 45 min and diffusional saturation at 600°C for 1 h.

(around large grains) on the autoradiogram correspond to the initial grain boundaries (Fig. 2a), while the light borders (around small grains) represent the structure of the recrystallized grains (Fig. 2b). The double network of boundaries is clearly visible on the autoradiogram in Fig. 3.

It is quite interesting that the network of the old boundaries on the autoradiogram is, first, incomplete, and, second, usually darker. The latter fact apparently results from the fact that the diffusion rate along the old boundaries after recrystallization at 700°C for 30 min is higher than the diffusion rate along the new boundaries of grains resulting from recrystallization.

Figure 4 shows the diffusion of carbon in samples recrystallized at 750°C for 1 h. The autoradiogram (Fig. 4c) shows one closed network of boundaries, which corresponds to the boundaries of new grains [shown in the micrograph (Fig. 4b)] and not to the grain boundaries of the deformed metal (before recrystallization).

Neither the micrographs or autoradiograms revealed any traces of the old grain boundaries.

The presence of tin in iron has a great effect on the diffusion of carbon in the metal.

This can be seen in that the temperature at which old boundaries disappear increases, and as the result the temperature at which the atoms begin to diffuse along the new boundaries is also higher. Thus, in pure iron recrystallized at 700°C for 30 min the carbon penetrates not only the old but also the new boundaries, while in iron containing tin the carbon does not diffuse at all along the new boundaries under the same conditions. Also, one can see considerable movement of carbon through the grains (Fig. 5).

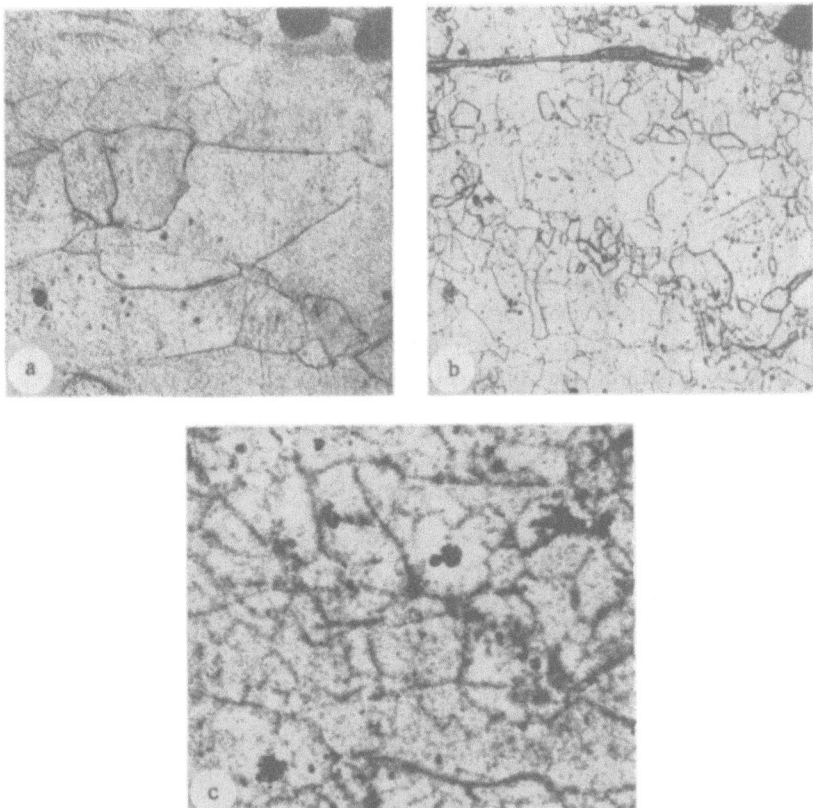

Fig. 2. Structure of iron. ×50. a) Microstructure after 56% deformation; b) microstructure after recrystallization at 700°C for 30 min; c) autoradiogram after recrystallization at 700°C for 30 min and diffusional saturation at 600°C for 1 h.

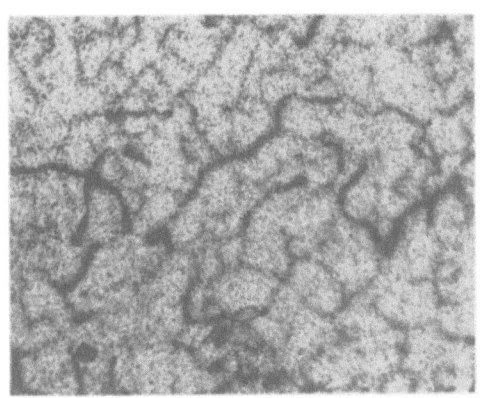

Fig. 3. Diffusion of carbon along the boundaries of the initial and recrystallized iron grains after recrystallization at 700°C for 30 min. Autoradiogram. ×50.

The double network of grain boundaries in iron containing tin occurs only after recrystallization at 750°C for 1 h (Fig. 6); in iron not containing tin it occurs after recrystallization at 700°C.

On the basis of photometric analysis of the autoradiograms (Fig. 7) we determined the dependence of the depth of diffusion along old and new boundaries on the recrystallization temperature.

The curves in Fig. 7 are partly experimental and partly arbitrary. Curves A and B refer to pure iron. Points 1 and 2 and points 4 and 5 are experimental, while points 3 and 6 are arbitrary. We were forced to use this method of drawing the curves for the following reasons. Points 2 and 5 on curves A and B were obtained from the same autoradiogram after recrystallization at 700°C (after which the double network occurs). Points 1 and 4, also obtained experimentally, correspond to the zero degree of darkening, i.e., the absence of old or new boundaries. Points 3 and 6, however, do not have a quantitative meaning, since the autoradiograms on which the degree of darkening was measured for these points were obtained under other experimental conditions than those corresponding to points 2 and 5, and therefore point 3 cannot be compared with point 2, and point 6 cannot be compared with point 5. However, the general direction of these curves indicates the character of the process.

Curves C and D, for recrystallized iron containing tin, were drawn in the same way. Thus, a direct investigation by the autoradiographic method showed that there is a singular "memory" or "inheritance" of the crystal with respect to the defects in the initial structure. Apparently this depends on the structure of the initial boundaries, i.e., defects and impurities, and also on the recrystallization conditions, i.e., the temperature and time. In iron this "memory" remains up to a temperature considerably higher than the recrystallization temperatures (650 and 700°C), while in iron containing tin it remains even at 750°C.

Consequently, the formation of a perfect structure in place of the old boundaries and the formation of a defective structure (the boundaries of new recrystallized grains) are processes which take a certain amount of time. These processes occur within a considerable temperature range and depend on a number of factors.

This result allows us to analyze a number of observations and to pose a number of theoretical questions.

Recrystallization is sometimes used to remove the effects of cold plastic deformation. However, according to our results, the residual effect may remain after the usual recrystallization process.

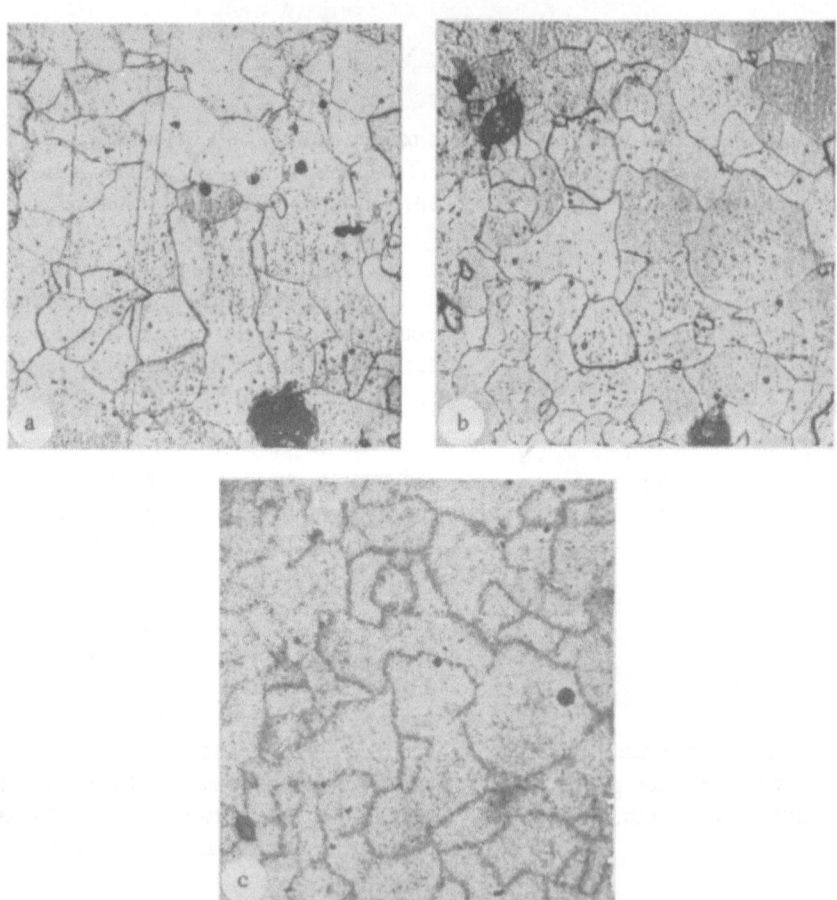

Fig. 4. Structure of iron. ×50. a) Microstructure after 54% deformation; b) microstructure after recrystallization at 750°C for 1 h; c) autoradiogram after recrystallization at 750°C for 1 h and diffusional saturation at 600°C for 1 h.

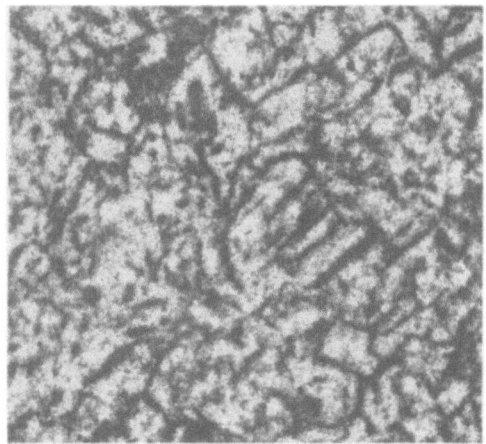

Fig. 5. Structure of iron containing tin after recrystallization at 700°C for 30 min and diffusional saturation at 600°C for 1 h. Autoradiogram. ×50.

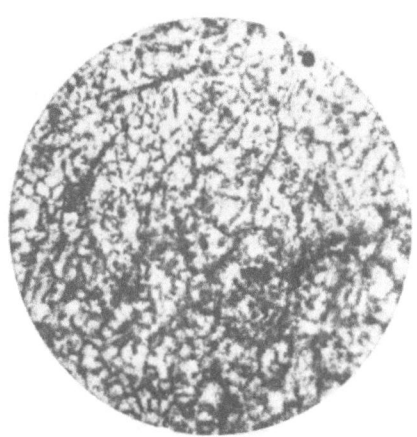

Fig. 6. Diffusion of carbon along the boundaries of initial and recrystallized grains of iron containing tin after recrystallization at 750°C for 1 h. ×30.

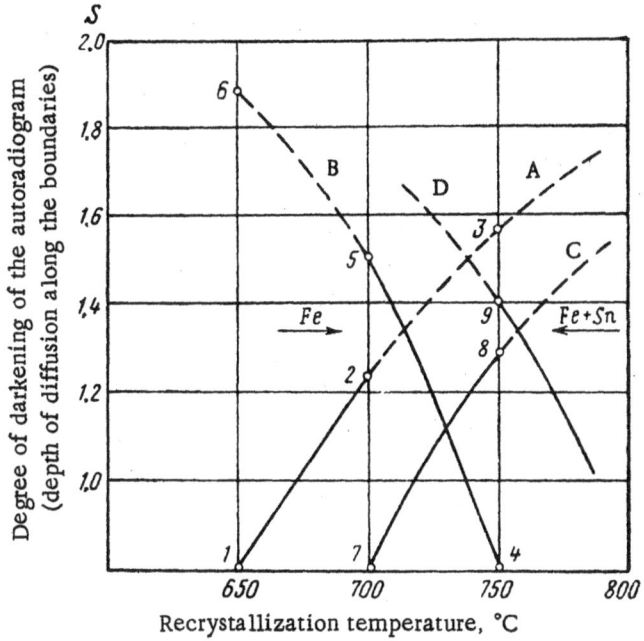

Fig. 7. Variation of the degree of darkening of old and new grain boundaries on the autoradiogram during recrystallization.

For example, it has been shown [1] that to obtain the same diffusional mobility in a plastically deformed sample as in an undeformed sample the deformed sample must be subjected to intermediate annealing at temperatures considerably higher than the recrystallization temperature.

It was shown that the diffusional mobility of atoms increases during the recrystallization process. Possibly this is due to the fact that diffusion occurs simultaneously along the old and new boundaries at some stage of recrystallization, and consequently the total length of the channels in which the mobility of atoms is high increases, particularly because recrystallization occurs at temperatures at which the boundaries play an important role in the diffusion process.

It is well known that recrystallization increases the creep rate of metals. This may also be due to the increased total length of the grain boundaries, when diffusion can occur simultaneously on the old and new boundaries at high temperature.

It is well known that at high temperatures the destruction of alloys can occur along the grain boundaries or through the grains. Possibly in this latter case the cracks follow the previously existing boundaries, which cannot be put in evidence metallographically.

The problem of the effect of impurities on the initial boundaries is then very important. Apparently, defects in the boundary usually slow down the formation of a perfect structure, particularly impurities with atoms very different from the atoms of the matrix.

It is quite probable that defects remain because of the presence of impurities and therefore the question arises whether it is possible to transfer the impurities from the boundaries into the grains by recrystallization.

The effect of harmful impurities could easily be eliminated if there were no "memory," but it is difficult to eliminate them from the boundaries of the original grains.

The results of the present investigation lead us to assume that the effect of harmful impurities can be decreased by recrystallization under certain conditions.

The results obtained here lead to an important theoretical and practical problem: Why, in the initial stages of the formation of new grains during recrystallization, do the impurity atoms diffuse along the old boundaries of new grains? It is clear that diffusion of impurity atoms and the destruction of interatomic bonds during etching of boundaries are very different methods of revealing defects in the structure.

It should be emphasized that on the autoradiograms with a double network of boundaries the old boundaries are much darker and are not closed.

Apparently, the healing of defects in some parts of the grain boundaries is more rapid than in others, possibly due to the irregular distribution of impurities along the boundaries or to the nonuniform structure of the boundaries themselves.

Thus, the imperfect structure of the boundaries of the initial grains is transformed into a perfect structure very gradually, but also irregularly, in a range of temperature and time.

LITERATURE CITED

1. S. Z. Bokshtein, T. I. Gudkova, A. A. Zhukhovitskii, and S. T. Kishkin, Some Problems of the Strength of Solids, Izd. Akad. Nauk SSSR, Vol. 76, 1959.

EFFECT OF INHERITED STRUCTURE IN MOLYBDENUM
AND MOLYBDENUM ALLOYS

S. Z. Bokshtein and M. B. Bronfin

The term "inherited structure" is defined as structural defects of the material in areas corresponding to the boundaries of grains existing before the growth of grains resulting from recrystallization. The existence of inherited structure was shown by an autoradiographic investigation of the diffusion of tungsten in molybdenum and molybdenum–zirconium alloys under the conditions described below.

The samples were annealed in vacuum at a temperature above the recrystallization temperature (1700°C for 14 h under a residual pressure of 10^{-3}-10^{-4} mm Hg). The surface of the samples was then coated electrolytically with a layer of W^{185}. The composition of the electrolyte and the conditions of electrolysis are described in the third article of this collection (p. 16). The activated samples were then annealed again at 1750°C for 108 h.

A comparison of the results from preliminary annealing and diffusional annealing shows that the boundaries of the grains migrate during diffusional annealing. According to [1], isothermal growth of grains in molybdenum at a temperature of the order of 1750°C terminates after 100 min and the grain size remains practically constant during the following 10^4 min. Thus, during most of the time of diffusional annealing of samples coated with radioactive tungsten the grain size is stable (1750°C).

Analysis of autoradiograms of samples cut at an angle to the surface coated with radioactive tungsten showed that tungsten diffuses not only along the grain boundaries stabilized at 1750°C but also along the original boundaries which existed before diffusional annealing. Figures 1a and 1b show, respectively, the autoradiogram and the micrograph of the same area of the surface of the sample of molybdenum containing 0.54% Zr. The comparison shows that W^{185} diffuses preferentially along both old and new grain boundaries. A separate section of this autoradiogram, shown in Fig. 2, confirms this result even more clearly.

These autoradiograms prove that under the conditions of diffusional annealing used some of the defects in the crystal lattice in the areas where the old grain boundaries existed before annealing remain after diffusional annealing. These defects in the areas of the old boundaries are apparently the reason for the greater diffusion in these areas of grains which have new separation boundaries.

The comparison between the degrees of darkening of the new and old boundaries on the autoradiograms at the same depth of diffusion shows that the concentration of the radioactive atoms is higher at the new boundaries than at the old boundaries. The greater concentration of impurity atoms on the new boundaries can be explained by the more intense diffusion along the new grain boundaries which, in turn, is related to a higher concentration of structural defects along the new boundaries than along the old boundaries. As the result, one observes double boundaries on all the autoradiograms of the materials investigated when the depth of diffusion is relatively low (0.2-0.3 of the maximum depth of diffusion of W^{185} on new boundaries).

It should be noted that the inherited structure is much clearer in molybdenum containing zirconium, apparently due to a higher stability of structural defects at the original grain boundaries resulting from the presence of oxides and carbides of zirconium, which are difficult to dissociate.

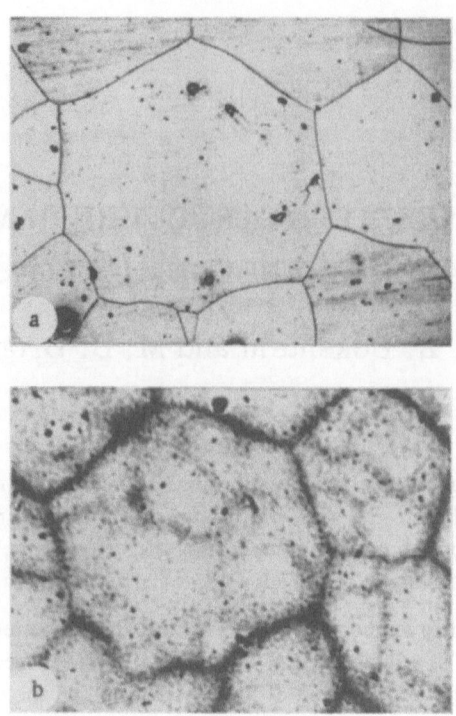

Fig. 1. Micrograph (a) and autoradiogram (b) of the same area of the surface of a sample of Mo+ 0.54% Zr coated electrolytically with W^{185} and subjected to diffusional annealing at 1750°C for 108. ×50.

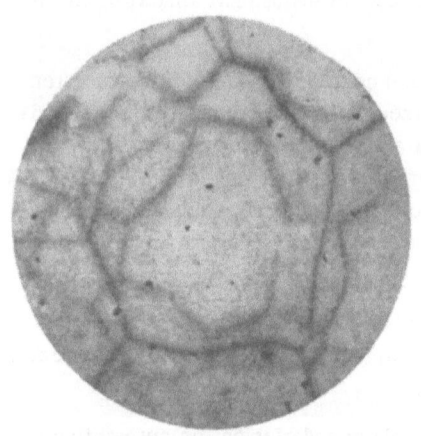

Fig. 2. Diffusion of W^{185} along the old and new grain boundaries in Mo+ 0.54% Zr at 1750°C. A section of the auto-radiogram in Fig. 1. ×30.

The fact that the structure remains loose in the area of the initial grain boundaries, as shown by the diffusional distribution of W^{185} in molybdenum, is a possible explanation of the effect of "vacuum etching" of old grain boundaries, which sometimes occurs during recrystallization of metals.

The possibility is not excluded that the looseness of the original grain boundaries is responsible to a certain degree for the high ductility of molybdenum at room temperature. This high ductility may be due to the redistribution of impurities inducing embrittlement, namely their transfer from the grain boundaries into the grains. One would expect that under certain recrystallization conditions the unhealed original boundaries could constitute traps for the embrittling impurity atoms (carbon, oxygen, and nitrogen), whose concentration along the newly formed boundaries becomes lower as a result. Such redistribution is probably possible during the initial stages of recrystallization, when the diffusional flow of penetration impurities with limited solubility in molybdenum lags behind the boundaries of the growing grains. Such a situation was observed experimentally in the recrystallization of iron containing radioactive carbon C^{14} along the original boundaries (see Fig. 13, p. 62).

LITERATURE CITED

1.   J. H. Bechtold, Trans. ASM 46: 1449, 1954.

EFFECT OF ULTRASONIC HEATING TO HIGH TEMPERATURES
ON THE GRAIN BOUNDARIES IN IRON

Yu. F. Balalaev and S. Z. Bokshtein

In practice, the structure of steel is often improved by recrystallization at temperatures above the Ac_3 point. The improvement in the structure is assumed to result from the formation of austenite out of a large number of nucleation centers. The problem of the relationship between the structure resulting from high-temperature heating and the initial structure is of great importance. In particular, it is important to determine whether the traces of the imperfect structure along the initial grain boundaries remain after recrystallization. The results obtained in [1], where the authors used both the autoradiographic and micrographic methods of analysis, seem to prove that recrystallization does not remove the inherited structure.

Usually, structural transformations in metals are induced by external application of heat or heating by electric current.

We investigated the behavior of grain boundaries during recrystallization annealing of iron with the following composition: 0.04% C, 0.07% Mn, 0.03% Si, 0.035% S, 0.015% P. The samples were heated to temperatures above 910°C by ultrasonic vibrations.

Ultrasonic vibration heats the steel up to 1000°C [2] by internal friction at the nodes of standing waves.

Iron rods were annealed at 950°C for 1 h to obtain a uniform small-grained structure. Samples 4 mm in diameter with a length equal to a half wavelength were cut from the rods. The samples were subjected to high-temperature heating by internal friction at the nodes of the standing ultrasonic wave, the frequency of oscillation being 19.5 kHz. For comparison, five samples were heated by electric current under identical conditions.

The samples were cut longitudinally along the center and examined under a microscope. The cold-hardened layer resulting from mechanical treatment and grinding was removed by electrolytic polishing in a Morris electrolyte [3] (133 ml CH_3COOH, 25 g CrO_3, 7 ml H_2O) at a temperature below 30°C, the current density being 2 A/cm^2. The samples were etched 20-30 sec, using a current density of 0.25 A/cm^2.

A temperature gradient occurs along the sample during ultrasonic heating as a result of sinusoidal distribution of stresses and the cooling of the ends of the samples. The existence of a temperature gradient over the length of the samples makes it possible to put in evidence the different structures in the same sample resulting from different temperatures.

In the present work we investigated the microstructure at different distances from the area of minimum temperature. Successive changes in the microstructure along the samples in the direction of increasing temperature are shown in the figure.

Ordinary metallographic analysis reveals grain boundaries with "crests" and "troughs."

The analysis of the microstructure shows that local etch pits begin to occur at the boundaries of old grains as the statistically average temperature of ultrasonic heating increases. These etch pits are not removed by recrystallization. No such etch pits occur after electric heating of the samples.

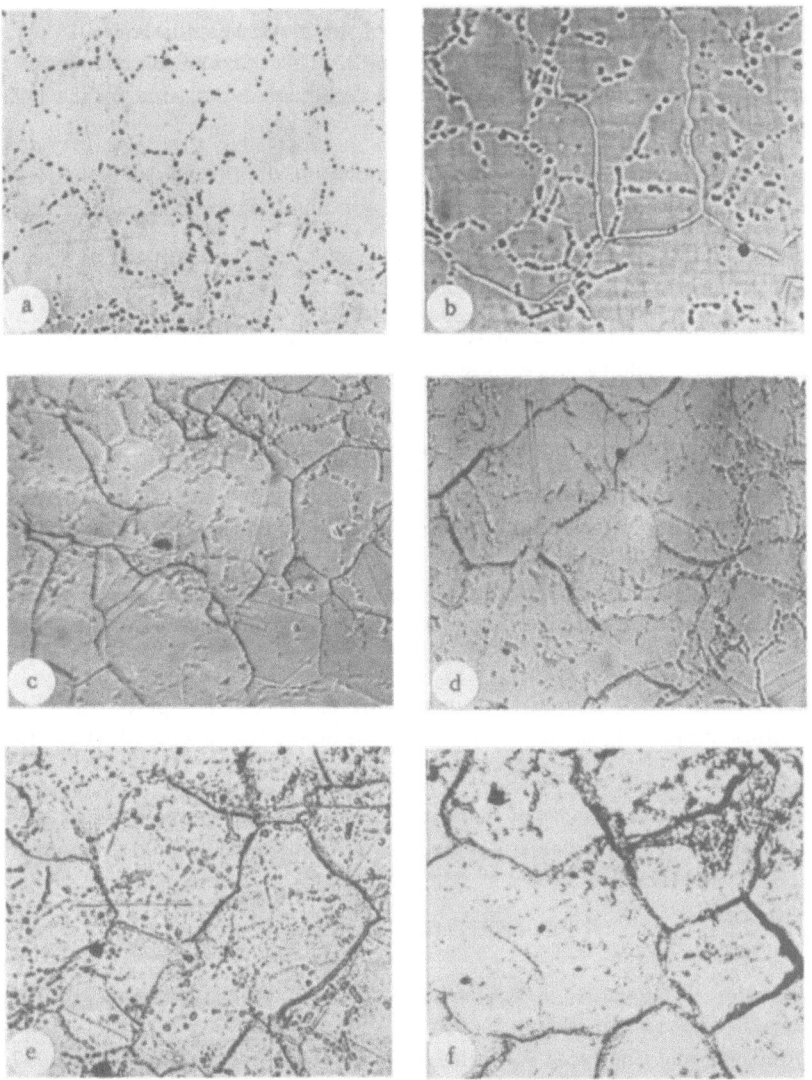

Microstructure of technically pure iron after ultrasonic heating to the maximum
temperature (1000°C) during 10 sec. The photographs show areas at distances of
12, 10, 8, 2, 1, and 0.1 mm from the fracture. ×270.

The occurrence of etch pits as the result of ultrasonic heating can be explained by internal friction resulting from defects in the grain boundaries. The relaxation of tangential stress along the surface of separation of phases and along the grain boundaries (characterized by large angles) is a well known source of internal friction in metals [4].

Relaxation processes along the grain boundaries occur in structurally imperfect microareas called "relaxation spheres."

Under the effect of ultrasonic vibrations a definite amount of thermal energy is evolved in the ductile relaxation spheres during each cycle. The total thermal energy induces heating of the metal to high temperatures, but the temperature of the ductile areas must be higher than that of the elastic matrix. Local thermal and deformational effects occurring during ultrasonic heating change the state of the relaxation spheres, which are thus revealed by etching. The change in the state of the relaxation spheres is apparently due to the concentration of vacancies and impurities. The relaxation spheres put in evidence as the result of ultrasonic heating do not overlap (see figure a): the relaxation in each sphere is independent.

Boundaries of new grains become visible in the area of the metal heated above 910°C; at the beginning these boundaries appear in the form of thin lines (see figure b). This shape of the boundaries indicates the absence of relaxation processes and small accumulations of the elements responsible for the lack of chemical uniformity between the boundary and the grain.

In areas heated to higher temperatures the etch pits along the old boundaries become less visible. New grains become larger with increasing temperatures and become much more visible (see figure c).

Thus, in the case of ultrasonic heating one observes not only the boundaries of new grains but also areas corresponding to the structural imperfections of old boundaries.

In some separate areas of new grain boundaries the structure is loosened, which indicates the beginning of viscous flow in the relaxation spheres located close together.

The etch pits along the boundaries of old grains gradually disappear, and this disappearance is promoted by high temperature, which eliminates structural and concentrational heterogeneities.

In the area close to the plane of destruction one observes only remnants of old grain boundaries. The boundaries of the new grains become loosened. The areas of the boundaries favorably oriented with respect to the direction of maximum tangential stresses, i.e., at a 45° angle to the longitudinal axis of the samples, the boundaries are the loosest (see figure d). In the area of the destruction one can see cracks; the final rupture of the sample at the end of ultrasonic heating occurs along the boundaries of the new grains formed during recrystallization (see figures e and f).

Thus, ultrasonic heating makes it possible to investigate the behavior of grain boundaries during recrystallization of iron. Ultrasonic heating reveals the most active relaxation spheres (ductile areas) on the grain boundaries.

This investigation revealed that traces of the imperfect structure of the boundaries of the original grains remain after recrystallization. At high temperatures the imperfections of the original structure gradually disappear.

LITERATURE CITED

1. S. Z. Bokshtein, M. A. Gubareva, S. T. Kishkin, and L. M. Moroz, Zavodskaya Lab., No. 10, 1960.
2. Yu. F. Balalaev, Zavodskaya Lab., No. 5, 1960.
3. V. Tegart, Electrolytic and Chemical Polishing of Metals [Russian translation], IL, 1957.
4. C. Zener, Elasticity and Inelasticity in Metals [Russian translation], 1954; A. S. Novik, Progress in the Physics of Metals, Vol. 1, Metallurgizdat, 1956.

STRUCTURAL SINGULARITIES AND DIFFUSIONAL MOBILITY

IN DIFFERENT PHASES OF TITANIUM ALLOYS

S. Z. Bokshtein, T. A. Emel'yanova,

S. T. Kishkin, and L. M. Mirskii

We have shown earlier [1] that in the case of self-diffusion or diffusion of different elements in metal alloys the atoms follow the grain boundaries preferentially. However, as was noted earlier, there is an exception to this rule. The diffusion of iron, chromium, and tin in titanium results in a special distribution of atoms: the atoms diffuse through the grain, but this diffusion is irregular and preferential along some surfaces of separation. The atoms diffuse preferentially along the boundaries only after 100 h of annealing at 750°C, which leads to the formation of a polyhedral α-structure. But even then there is a certain amount of irregular diffusion through the grains.

On the basis of these results, we have assumed that the polymorphic transformation induces a specific structural state in titanium alloys and that in this state the grains possess a large number of channels in which the atoms have a high diffusional mobility.

In this article we describe the results of an investigation made on this assumption.

Our main problems were to put in evidence the structural singularities resulting from the $\beta \rightarrow \alpha$ transformation and also to measure the depth of diffusion in different phases of titanium alloys.

Since carbon diffuses preferentially along the grain boundaries of α- and γ-iron [2], it can be assumed that the carbon atoms follow essentially the defects in the structure of the crystal. Thus, we investigated the structural singularities of titanium alloys by studying the diffusion of carbon, and in some cases nickel, in the β-, α-, and α'-phases.

The chemical composition of the alloys investigated is given in Table 1.

The structural singularities resulting from polymorphic transformation were studied in a Ti—Mn alloy and also the VT3-1 alloy.

The depth of diffusion in the β-and $(\alpha + \beta)$-phases was studied by the diffusion of molybdenum in the VT15 alloy.

And finally, the VT3-1 alloy was used to study the diffusional mobility in a thin surface layer containing the α-component either in the shape of needles or polyhedrons.

TABLE 1

Alloy	Basic elements, %						Impurities, %		
	Ti	Mn	Al	Cr	Mo	Fe	C	Si	Fe
Ti+3.83% Mn	Base metal	3.83	—	—	—	—	0.038	—	0.16
VT3-1	" "	—	4.9	1.8	1.75	0.45	0.11	0.06	—
VT15	" "	—	3.69	9.43	8.08	—	—	—	—

Fig. 1. Distribution of carbon in the Ti+3.83% Mn alloy after annealing 7 h at 950°C and cooling in the furnace. ×50. a) Micrograph; b) autoradiogram of the same area.

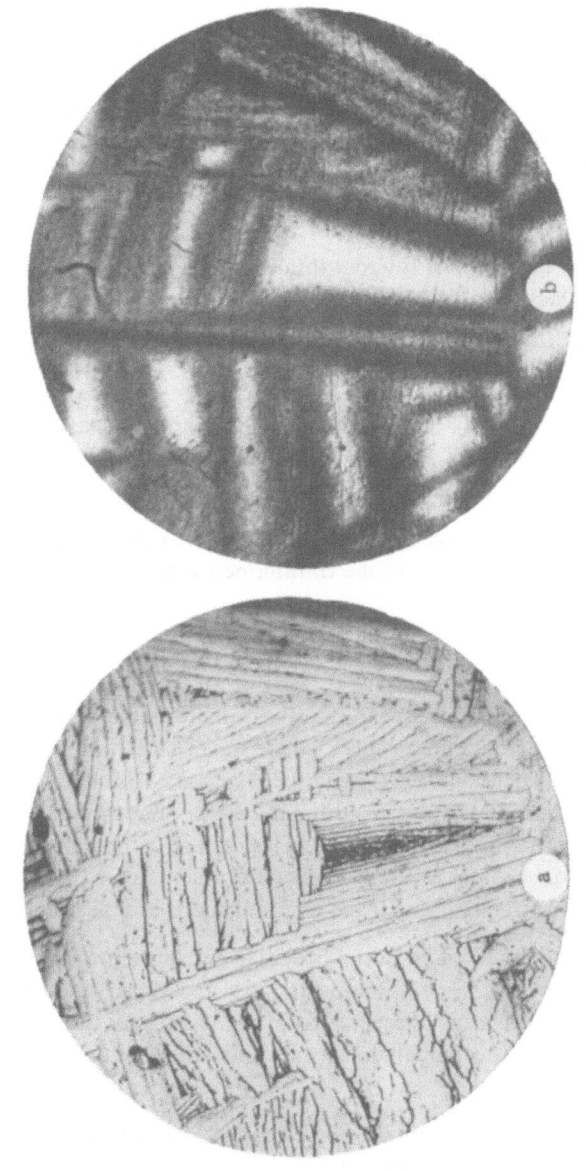

Fig. 2. Distribution of carbon in the Ti+3.83% Mn alloy after annealing at 1050°C for 10 h and cooling in the furnace (preliminary saturation with carbon at 750°C for 20 h). ×50. a) Micrograph; b) autoradiogram of the same area.

INVESTIGATION OF THE STRUCTURAL SINGULARITIES OF DIFFERENT PHASES

OF TI+3.83% Mn AND THE VT3-1 ALLOY

We used the autoradiographic method to study the distribution of elements resulting from diffusion [3]. The radioactive isotopes used were C^{14} and Ni^{63}.

The Ti—Mn alloy was annealed 4 h at 1000°C in vacuum and cooled in the furnace. This treatment produced a structure of large needles oriented in the direction of the large grains of the initial β-phase. Then the samples were saturated with carbon from barium carbonate containing radioactive carbon. The samples were then subjected to cementation at 950 and 1050°C in evacuated (10^{-2} mm Hg) quartz ampules. After heating at 950 and 1050°C for 7-10 h the samples were quenched, or cooled in the furnace, or subjected to isothermal annealing in the (α + β)-region (780°C). The samples were quenched by immersing the ampule containing the samples in a 10% solution of sodium hydroxide.

The samples of the VT3-1 alloy were first quenched from 1100°C, then coated electrolytically with Ni^{63}, and then subjected to diffusional annealing at 850°C for 50 h.

After this treatment a very thin layer (a few microns thick) was removed from the samples by mechanical polishing and the samples were placed on NIKFI photographic plates of the MR type.

The autoradiograms were compared with micrographs of the same area of the samples. The results and a discussion of the results are given below.

The autoradiograms of the samples of Ti+3.83% Mn quenched from 950°C after cementation showed that carbon is uniformly distributed in the single phase β-region and that there is no significant segregation along the grain boundaries.

Slow cooling in the furnace from 950°C produces a redistribution of carbon, as shown in Fig. 1. The carbon atoms move preferentially to the boundaries of the transformed β-phase and to the surfaces of separation between the largest needles of the α'-phase.

Such an irregular distribution of carbon cannot be the result of preferential diffusion along the inherited defects formed earlier in the crystal lattice of the β-phase. First of all, the grains of the β-phase and the needles of the α'-phase on the autoradiogram correspond exactly to those on the micrograph of samples heated to 950°C and cooled slowly. The β-grains and the α'-needles are much larger after heating at 1000°C. Secondly, quenching produces a regular distribution of carbon in the β-phase. Thirdly, to heal any inherited defects in the crystal lattice, a series of samples saturated with carbon was subjected to 20 h annealing at 750°C and 10 h at 1050°C and then cooled in the furnace. In this case, also, the carbon atoms migrate to the separation boundaries of the α'-phase, but the needles of the α'-phase are much larger (Fig. 2). Consequently, the segregation of carbon atoms occurs during cooling and is the result of the β →α' transformation.

We believe that the only possible explanation of the irregular distribution of carbon during cooling is the peculiar (martensitic) mechanism of the β→ α' transformation. Apparently, during this transformation the crystal lattice becomes irregular at the boundary of separation of the needles of the α'-phase. This local deformation of the crystal lattice is probably accompanied by mass movement of atoms which have penetrated the lattice.

It is quite interesting that the carbon atoms move only to the surfaces of separation between the largest needles of the α'-phase, which form at the highest temperature. With increasing temperatures the space between the large needles becomes filled with small precipitates. Apparently, the redistribution of carbon during cooling is the result of acts of diffusion which cannot occur at low temperatures.

Similar results were obtained in a work concerning the distribution of nickel during the diffusion of nickel in titanium [4]. Quenching resulted in regular distribution of nickel in the β-region, while slower cooling produced accumulations of nickel atoms on the surfaces of separation of the α'-phase.

Figure 3 shows an autoradiogram of the Ti—Mn alloy heated at 1050°C, cooled to 780°C, and kept at this temperature for 50 h, and then cooled rapidly. At 780°C a binary phase is formed: (α + β). The figure

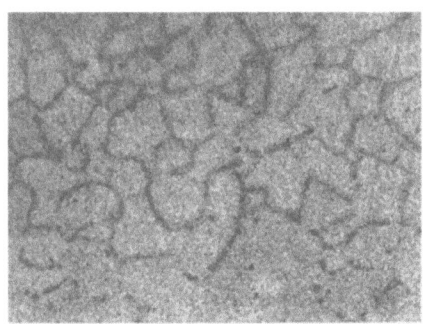

Fig. 3. Distribution of carbon in Ti + 3.83% Mn after isothermal annealing under the following conditions: heated 18 h at 1050°C, cooled to 780°C and kept at this temperature 50 h. Autoradiogram. ×50.

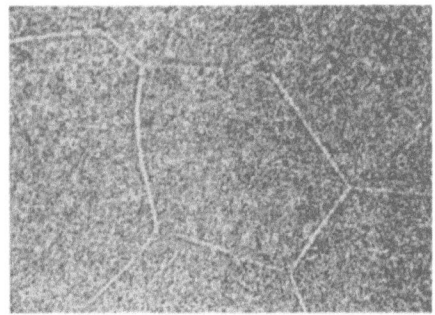

Fig. 4. Distribution of nickel resulting from diffusion in the α'-phase of the VT3-1 alloy at 850°C. Autoradiogram. ×50.

shows that the distribution of carbon during isothermal heating is very different from the distribution resulting from continuous cooling. The carbon atoms are uniformly distributed through the grain and there is preferential diffusion in the α-phase, which is located along the β-grain boundaries. Apparently, isothermal heating at 780°C leads to an equilibrium β → α transformation, to which this distribution corresponds.

The diffusion of nickel directly in the α'-phase agrees with the previous result. The autoradiogram in Fig. 4 shows irregular distribution of nickel resulting from diffusion in the α'-phase of the VT3-1 alloy. Nickel diffuses preferentially along the surfaces of separation between the needles of the α'-phase. The smaller the needles the more uniform the darkening of the autoradiogram. This distribution of nickel confirms the presence of greatly deformed regions between the needles of the martensite phase. These regions are the most favorable channels for the diffusion not only of penetrated atoms but also for the atoms of the element forming a substitution solid solution with titanium.

Comparison of Fig. 1b and Fig. 4 shows that the grain boundaries of the transformed β-phase appear black on the autoradiogram in the case of diffusion of carbon and light in the case of diffusion of nickel. The largest needles of the α-phase are formed on the boundaries of the original β-grains. It is quite possible that these needles are formed not according to martensitic kinetics but according to isothermal kinetics. Therefore, the difference in the autoradiograms in Fig. 1b and Fig. 4 can be explained by the difference in the diffusional mobility of carbon and nickel in the equilibrium α-phase. Carbon, being an α-stabilizer, dissolves easily in the α-phase, while the diffusional mobility of nickel is low in this phase because the solubility of nickel in α-titanium is very low.

DIFFUSIONAL MOBILITIES OF MOLYBDENUM IN THE HETEROPHASE
AND SINGLE-PHASE REGIONS OF THE VT15 ALLOY

We studied the depth of diffusion of Mo^{99} in the α- and (α + β)-regions of the VT15 alloy.

The diffusion parameters were determined by the thin-layer method [5].

Experiments using radioactive molybdenum are rather difficult. The relatively short half-life of this isotope (only 68 h) requires short experiments and a high initial activity of the sample. On the other hand, because of the very hard radiation, the diffusional heating must be very long in the case of the absorptional method in order to obtain any significant change in the activity. To investigate the diffusional mobility of molybdenum in the heterophase region of the alloy, the diffusional annealing temperature must be below the temperature of the polymorphic transformation, i.e., below 750°C. At these low temperatures the diffusional mobility is very low, which also makes the investigation rather difficult.

We cut rectangular samples 10 × 15 mm out of foil 95 μ thick.

The preliminary heat treatment consisted of annealing some of the samples at 600°C to obtain the heterophase (α + β)-structure and quenching other samples from 850°C to obtain a single-phase β-structure.

TABLE 2

Temperature of diffusional annealing, °C	Phase in VT15	Diffusion coefficient of molybdenum, $D \cdot 10^{11}$, cm^2/sec
700	$(\alpha + \beta)$	0.13
875	β	1.55
900	β	2.58
925	β	3.81
1000	β	11.04

Then one side of the sample was electrolytically coated with a layer of Mo^{99} and the sample subjected to diffusional annealing, and after this the activity was counted on both sides of the foil. The samples with a heterophase structure were annealed at 700°C, while the samples with a single-phase β-structure were annealed at 875, 900, 925, and 1000°C. The samples were annealed in vacuum (residual pressure, 10^{-5} mm Hg) in quartz tubes which were inserted in a tubular electric furnace. The temperature was controlled with a chromel—alumel thermocouple and was regulated automatically with an EPD-09 potentiometer with a precision of ±5°C. The samples were annealed 100 h at 700°C and 40 h at 875-1000°C. The activity of the samples was calculated in the first case after every 50 h, and after every 4 h in the second case. The heating and cooling time did not exceed 3 min.

The activity was measured with a radiometric apparatus of the B-2 type. The results of the investigation are given in Table 2.

The temperature dependence of the diffusion coefficient of molybdenum in the VT15 alloy is shown in Fig. 5.

In the β-region the experimental points follow a straight line quite well; we determined the value of the activation energy of the process from the slope of the line. According to our calculations, this value is 45,000 cal/g-atom.

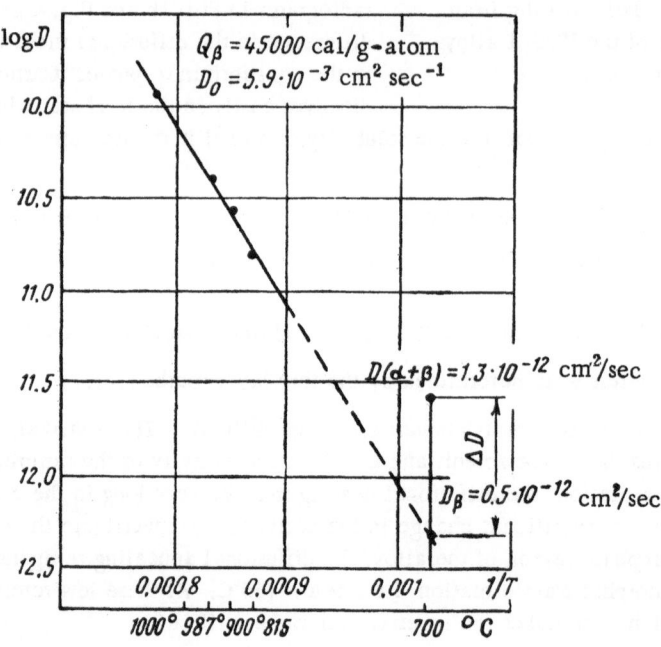

Fig. 5. Temperature dependence of the diffusion coefficient of molybdenum in the β-phase of the VT15 alloy.

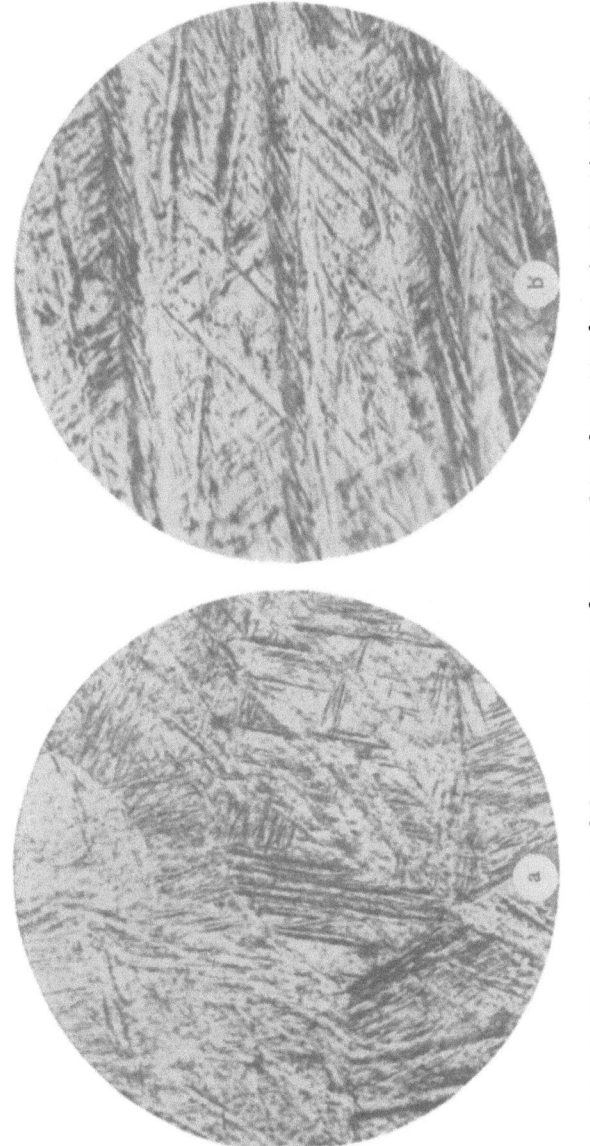

Fig. 6. Microstructure of the VT3-1 alloy after quenching from 1100°C. a) ×100; b) ×500.

89

Fig. 7. Microstructure of the VT3-1 alloy after cold deformation, annealing, and phase recrystallization. a) ×100; b) ×500.

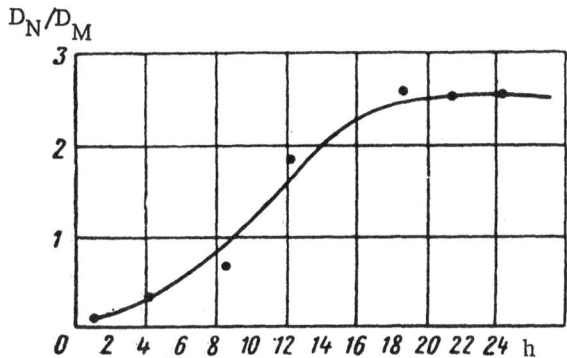

Fig. 8. Variation of the ratio of the coefficients of surface diffusion of nickel in the needle and massive α-phases of the VT3-1
alloy with the annealing time.

In the same figure we show the value of the diffusion coefficient of molybdenum in a two-phase region
at 700°C, and by extrapolation we obtain a point corresponding to diffusion in the β-phase at 700°C. The figure shows that the diffusional mobility of molybdenum in the heterophase region is about 2.7 times greater than
in the single-phase β-region.

These results can be explained by the higher rate of diffusion of molybdenum into the α-phase. Possibly,
the diffusional mobility of molybdenum in titanium is similar to the diffusional mobility of tin. It was shown
in [6] that the diffusion coefficient of tin in α-titanium is much higher than in β-titanium. The amount of
the α-phase in the alloy we investigated was about 25% at 700°C.

However, another no less probable explanation is the preferential diffusion of molybdenum along the
boundaries of separation of phases.

An additional investigation is needed to find a solution to this problem.

CHANGE IN THE DIFFUSIONAL MOBILITY OF NICKEL IN THE SURFACE LAYER

OF THE VT3-1 ALLOY WITH THE SHAPE OF THE α-PHASE

In this investigation we used Ni^{63}. The diffusion coefficients in the surface layer were determined by the
absorption method. The absorption coefficient was determined experimentally in VT15 foil and was found to
be equal to 336 cm^{-1}.

A series of samples was quenched from 1100°C, which led to the formation of the α'-phase (Fig. 6).

The second series of samples was cold-worked (deformed 8%), annealed 3 h at 800°C, and then subjected
to phase recrystallization (a short time at 1000°C, cooled to 800°C and kept at this temperature 1 h, and cooled
in the furnace to room temperature). This treatment led to the formation of a massive α-phase which was not
completely equiaxial and was very different from the needle-shaped α-phase (Fig. 7).

After this treatment both series of samples were coated electrolytically with nickel and subjected to diffusional annealing at 800°C in vacuum. The samples of the needle-shaped phase and the massive α-phase
were annealed for seven different periods of time, varying from 1 to 4 h, side by side. After each period of
annealing the activity of the samples was measured; the diffusion coefficient in the surface layer was determined by the decrease in the activity during a given period of time.

To determine the effect of the shape of the α-component on the diffusion rate, the ratio between the
diffusion coefficient of nickel in the needle-shaped α-phase and in the massive α-phase (D_N/D_M) was determined after each annealing period. The variation of this ratio as a function of the annealing time is shown
in Fig. 8.

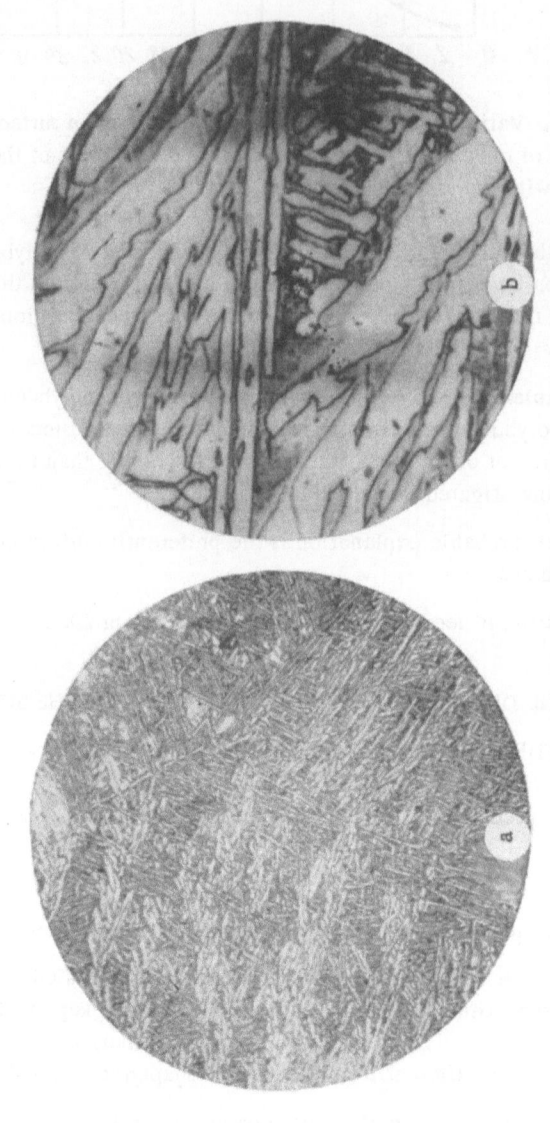

Fig. 9. Microstructure of the VT3-1 alloy with an initial needle α-phase after diffusional annealing at 800°C for 24.5 h. a) ×100; b) ×500.

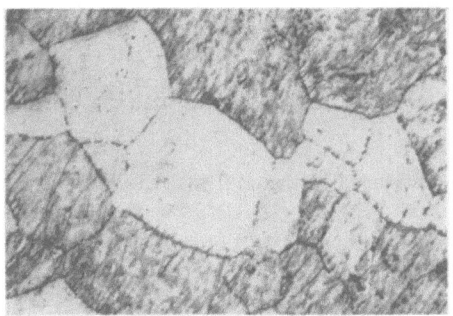

Fig. 10. Microstructure of the VT3-1 alloy with an initial massive α-phase after diffusional annealing at 800°C for 24.5 h. ×100.

The D_N/D_M ratio increases during the initial stages of annealing and later (after 15-18 h) becomes constant, D_N being greater than D_M.

These results can be explained as follows. Quenching from 1100°C leads to a martensite-type transformation with the formation of the α'-phase, which during annealing is transformed into the needle-shaped α-phase, which is very stable during heating. The micrograph (Fig. 9) shows that diffusional annealing at 800°C for 24.5 h does not lead to recrystallization of the needle-shaped phase but induces only coagulation of a few needles. Apparently recrystallization is prevented by the boundaries of separation between needles, to which the impurity atoms move.

Apparently these boundaries are destroyed during cold deformation, annealing, and phase recrystallization, and conditions favorable for recrystallization are created (Fig. 10). Recrystallization accompanied by an increase of diffusional mobility occurs during the first stages of annealing in these samples; this does not occur in samples with a needle-like structure. Therefore, the permeability of the massive α-phase is higher at the beginning of diffusional annealing. After the equiaxial grains are formed, the diffusional mobility of nickel in these grains is lower by a factor of 2.5 than in the needle-like α-phase. This is due to the fact that the needle-like α-phase has a greater number of surfaces of separation and areas in which the crystal lattice is disrupted and diffusion proceeds essentially in these areas.

These results confirm the conclusion arrived at in a number of previous investigations concerning the increase of diffusional mobility during recrystallization and phase recrystallization processes and also the role of interphase separation surfaces during diffusion.

CONCLUSION

Our investigation showed the existence of some structural singularities occurring in titanium alloys during the $\beta \rightarrow \alpha$ transformation. When this transformation obeys martensitic kinetics the crystal lattice at the surfaces of separation of the α'-phase is greatly disrupted, and this leads to an increase in diffusional mobility. These disruptions, which are stable during further heating, undoubtedly have an effect on the mechanical properties of titanium alloys.

The investigation of the distribution of carbon and nickel in titanium alloys showed the following:

a) in the case of diffusion into the β-region of the Ti—Mn alloy, the carbon is distributed uniformly without any apparent segregation along the grain boundaries;

b) during continuous cooling of this alloy carbon atoms move rapidly from the β-region to the separation boundary of the α'-phase;

c) after isothermal transformation of the β-phase, with the formation of equiaxial α-phase (plus residual β-phase), the carbon is uniformly distributed through the grain and diffuses preferentially into the α-phase;

d) when nickel diffuses directly into the α'-phase of the VT3-1 alloy the atoms accumulate at the surfaces of separation of needles of this phase. Nickel diffuses preferentially along the grain boundaries only after prolonged annealing at a temperature below the temperature at which the polyhedral α-phase is formed. However, even in this case there is still heterogeneous diffusion through the grains, which indicates that the crystal lattice is disrupted within the grain.

Comparison of the diffusional mobilities in different phases of the titanium alloy showed that:

a) the mobility of molybdenum diffusing in the β-phase of the VT15 alloy is lower than in the heterophase region of the same alloy;

b) the diffusional permeability of the equiaxial α-phase in the VT3-1 alloy is lower than the diffusional permeability of the needle-shaped α-phase.

Recrystallization of the α-phase in the VT3-1 alloy leads to increased diffusional mobility of atoms.

LITERATURE CITED

1. S. Z. Bokshtein, S. T. Kishkin, and L. M. Moroz, Investigation of the Structure of Metals by the Radioactive Isotope Method, Oborongiz, 1959.
2. S. Z. Bokshtein, M. A. Gubareva, I. E. Kantorovich, and L. M. Moroz, Metalloved. i Term. Obrabotka Metal.,No. 1, 1961.
3. S. Z. Bokshtein, S. T. Kishkin, and L. M. Moroz, Investigation of the Structure of Metals by the Radioactive Isotope Method, Oborongiz, 1959; M. E. Drits, Z. A. Sviderskaya, and É. S. Kadaner, Autoradiograms in Metal Science, Metallurgizdat, 1961.
4. R. F. Peart and D. H. Tomlin, Acta Met. 10 (2): 123-134.
5. S. N. Kryukov and A. A. Zhukhovitskii, Dokl. Akad. Nauk 90: 379, 1953.
6. S. Z. Bokshtein, S. T. Kishkin, and V. B. Osvenskii, Metalloved. i Term. Obrabokta Metal.,No. 1, 1961.

EFFECT OF STRESS AND STRUCTURE ON PORE FORMATION
AND THE DESTRUCTION OF METALS AT HIGH TEMPERATURES

A. A. Blistanov, S. Z. Bokshtein, T. I. Gudkova, S. T. Kishkin,
and A. A. Zhukhovitskii

The destruction of metals at high temperatures is a complex phenomenon and there are relatively few studies of the heat resistance of metals.

The displacement of atoms and dislocations by diffusion at high temperatures resulting from external stress leads, under certain conditions, to the formation of vacancies in the crystal lattice, and the concentration of vacancies becomes greater than the equilibrium concentration.

The process of pore formation is very interesting because it relates the initial microscopic stage of the formation of defects in the crystal lattice to the final macroscopic stage of destruction.

A pore is a link between an atomic hole (occurring as the result of elementary acts of diffusion and displacement of dislocations under the effect of temperature fields or stress fields) and the destruction of the material.

Some investigators at home and abroad [1] have studied the effect of different factors on the process of pore formation, but all these investigations concern essentially the qualitative characteristics of pore formation.

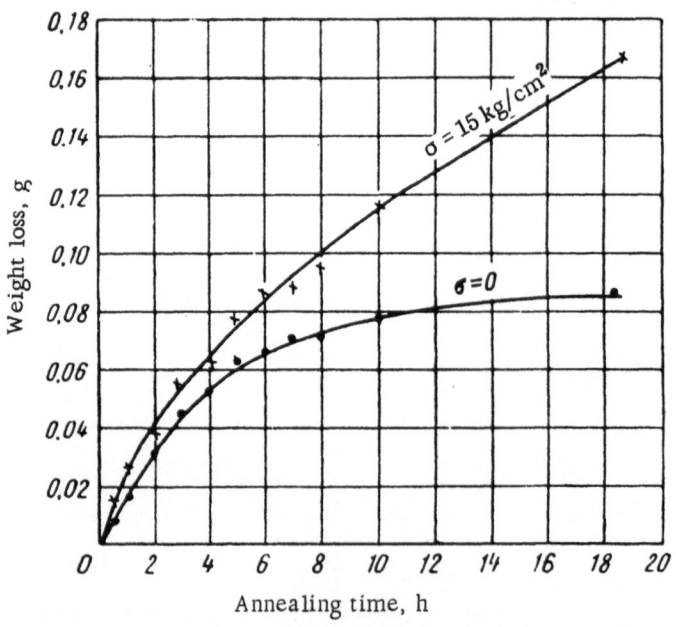

Fig. 1. Effect of stress on the weight loss of the sample during diffusional annealing at 750°C.

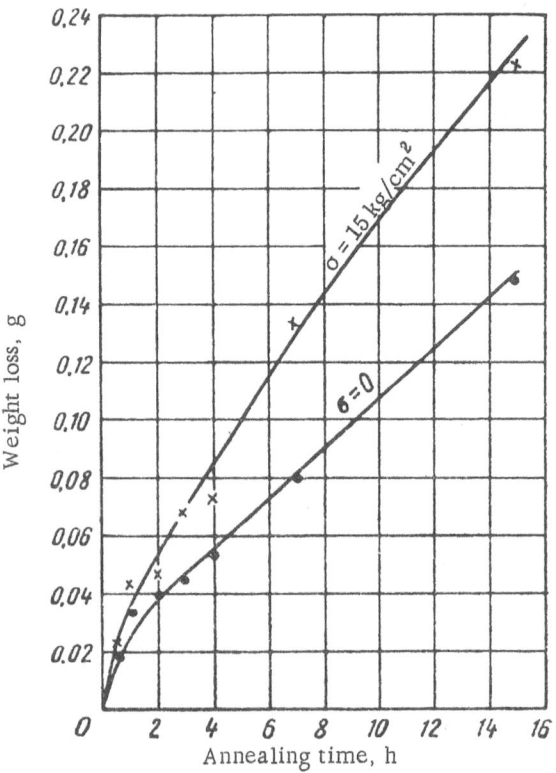

Fig. 2. Effect of stress on the weight loss during diffusional annealing at 800°C.

We investigated the effect of stress on the process of pore formation. Great attention was also given to the effect of the structure of the metal.

EXPERIMENTAL METHOD

Our experiments were made with α-brass composed of 32% zinc and 68% copper and nichrome composed of 20% chromium and 80% nickel.

Fig. 3. Microstructure of brass after degassing at 800°C for 2 h. ×100.

Since the vapor pressures of zinc and chromium in these alloys are higher than those of copper and nickel, the concentration of vacancies necessary for the formation of pores was produced by degassing the samples in vacuum at high temperatures. It should be noted that brass of this composition does not undergo any phase transformation up to the melting point. Brass samples were cut from a sheet 2 mm thick and annealed in quartz ampules 50 h at 800°C under a pressure of 10^{-2} mm Hg. Under these conditions the stress is completely relieved and the grains grow.

After annealing, the samples were electropolished in an electrolyte consisting of orthophosphoric acid with a current density of 1.55 A/cm^2. A layer up to 15 μ thick was removed by 1 h of electropolishing.

The nichrome ingots were forged and rolled into sheets 5 mm thick. Samples were prepared from these sheets and subjected to homogenizing annealing in argon at 1200°C for 50 h. The annealing relieved the stresses and increased the grain size.

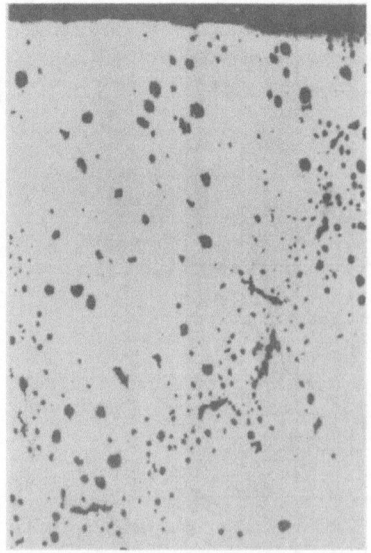 

Fig. 4. Microstructure of brass after degassing at 800°C for 15 h. ×100.

Fig. 5. Microstructure of brass after degassing at 800°C for 2 h under a stress of 15 kg/cm². ×100.

After annealing, the samples were electropolished in an electrolyte consisting of 1000 ml of acetic acid and 50 ml of chloric acid.

Diffusional annealing was done in a special apparatus in which the volatile component was evacuated while the samples were subjected to different tensile stresses. The residual pressure was 10^{-3} mm Hg and the temperature was varied between 500 and 1250°C. The applied stress was varied from 15 to 120 kg/cm². The highest stress was used at the lowest temperature, and vice versa.

Two samples were annealed simultaneously in all cases. One was subjected to tensile stress and the other was suspended freely.

The samples were etched for microstructural analysis in the same bath as that used for electropolishing, the only difference being that the current density was lower. The voltage used was 0.4–0.6 V for brass and 25 V for nichrome.

The α-brass samples were cut at an angle to the surface to study the distribution of pores in the grains; this method makes it possible to determine the distribution of pores through a section of the samples no thicker than one grain.

The process of pore formation was studied quantitatively by measuring the variation of the density of the samples by the hydrostatic method.

The samples were weighed on an analytic balance with a precision of 0.0002 g.

The precision in the determination of the volume of the samples was 0.005 cm³. Diffusional annealing produces small cracks and pores on the surface of the samples, and these have a great effect on the density. The annealed samples were rubbed with oil to eliminate these errors.

RESULTS

Kinetics of Pore Formation and the Effect of Stress

Samples of α-brass were subjected to diffusional annealing in a vacuum apparatus at 750 and 800°C. One of the samples was unstressed and the other was under a tensile stress of 15 kg/cm². The tensile stress was increased to 30 and 40 kg/cm² at later stages of the test. The samples degassed at 750°C were weighed after

0.5, 1, 2, 3, 4, 5, 6, 7, 8, 10, and 18.5 h; the samples degassed at 800°C were weighed after 0.5, 1, 2, 3, 4, 7, 10, and 15 h. The periods of 18.5 and 15 h correspond to the moments of the destruction of the samples at the respective diffusional annealing temperatures and stresses.

The results are shown in Figs. 1 and 2.

The results show that at 750 and 800°C much more zinc evaporates from samples under stress than from samples not stressed, apparently due to the effect of stress on the diffusion rate.

The effect of stress on evaporation is particularly strong during the initial periods of testing and in the last stages preceding destruction.

With increasing diffusional annealing temperatures, the effect of stress on the rate of evaporation decreases and is significant only in the last stages preceding destruction. Thus, at 800°C the weight loss of the ruptured samples is about 1.5 times higher than that of the unstressed samples (0.2204 g and 0.1479 g, respectively).

The decrease of the effect of stress on evaporation at high temperatures can be explained by the increase in the diffusional mobility of atoms, as was shown in previous experiments [2].

Figures 3-7 show the microstructure of samples after evaporation at 800°C for different times. It should be noted that there are no pores in samples tested for 0.5 h, apparently due to the fact that during the early stages of the evaporation process all the additional vacancies migrate to the surface, or to the fact that the size of pores is so small that they cannot be seen under the magnification used in this investigation. The number of pores increases with the time of diffusional annealing. The small pores begin to coagulate and deeper areas become porous.

The process of coagulation is accompanied by the formation of small pores in areas further removed from the surface of the sample. After diffusional annealing for 15 h the pores reach a considerable depth (up to 800 μ). The newly formed small pores are concentrated essentially along the grain boundaries and in the immediate vicinity. The coagulation of large pores leads to the weakening of the grain boundaries and to the formation of small fissures (see Fig. 4). The fissures are preceded by the formation of pores.

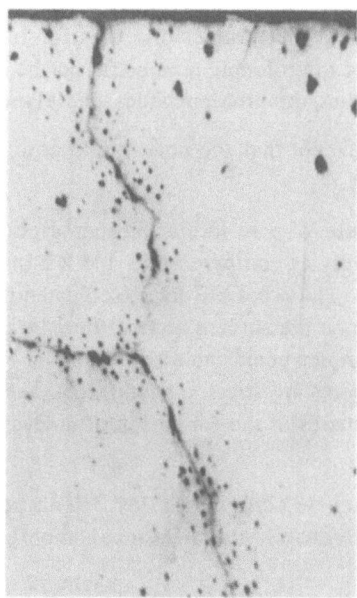

Fig. 6. Microstructure of brass after degassing at 800°C for 10 h under a stress of 15 kg/cm². ×100.

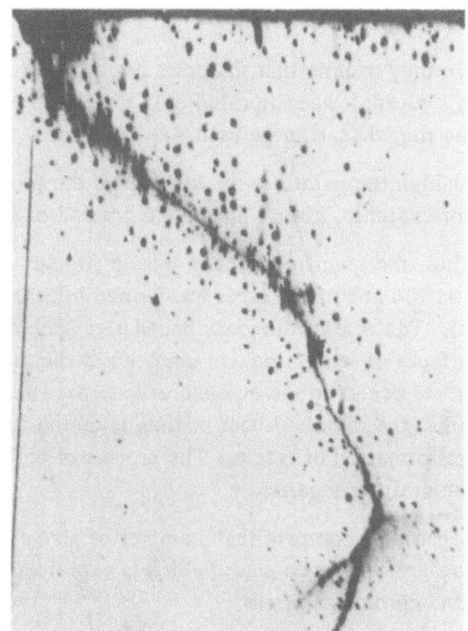

Fig. 7. Microstructure of brass after degassing at 800°C for 15 h under a stress of 15 kg/cm². ×100. The sample was destroyed.

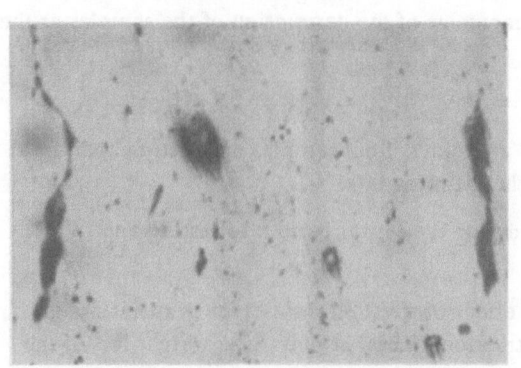

Fig. 8. Microstructure of nichrome after degassing at 1200°C for 2 h and 15 min under a stress of 70 kg/cm². ×300.

The mechanism of the formation and development of pores in the stressed samples is somewhat different (Figs. 5-7). After degassing for 1 h, pits are formed on the surface of the sample; these pits are places of the preferential evaporation of zinc. With increasing degassing time pores and then cracks begin to develop along the grain boundaries (Figs. 6 and 7).

It should be noted that if the grain boundary at a certain distance from the surface is weakened by one means or another then pores begin to form within the metal close to this defective boundary. In this case the cracks may not begin at the surface and may form either on the surface or within the metal.

During later stages of the test, when the grain boundaries are already weak, the volatile component is evacuated from the grain boundaries in all cases through a channel which opens to the surface of the sample (Fig. 6). In this case large numbers of small cracks form on the grain boundaries. These cracks unite and form a crack which spreads into the sample to a considerable depth.

The volatile component evaporates intensely along the grain boundaries in areas next to the fracture of the sample. The small pores coagulate into large pores, which combine into a crack along the grain boundary. The channel along which the evaporating component passes becomes wider and deeper and this, in turn, increases the rate of evaporation from the grain boundaries (Fig. 7).

The analysis of the kinetics of this process leads to the following conclusions about the mechanism of the process. The formation of pores is initiated as the result of evaporation and the simultaneous diffusional displacement of the evaporating component. The grain boundaries have no clear effect in the initial moments of the process. Then pores begin to form. Thorough analysis shows that the pores occur first of all close to the boundaries; in separate cases small cracks can be seen in these boundaries, which obviously leads to weakening of the boundaries.

We may assume that the pores are also formed more intensely at the surfaces of separation in unstressed samples. We must keep in mind only that large numbers of pores of microscopic size could not be seen because the magnification we used was insufficient to reveal them. Thus, this problem needs special investigation.

At high temperatures the kinetics of the process is the same except that the pores begin at an earlier stage (for example, after 2 h at 800°C instead of after 3 h at 750°C).

Thus, the specific effect of tensile stress is as follows. The rate of pore formation increases (for example, at 750 and 850°C pores are formed after 2 and 1 h, respectively, as compared to 3 and 2 h in unstressed samples). The role of the grain boundaries becomes much greater. The pores are localized essentially on the inner surfaces of separation. In many cases the pores begin to form on the surface around the edges of the pits, which never occurs in the absence of external stress. In stressed samples pores can also occur along the boundaries within the metal without beginning on the surface. Pits and pores are stress concentrators, and therefore favor the formation of cracks. The process of pore formation terminates in the formation of cracks and fissures along the grain boundaries.

Let us now compare the character of pore formation in the nickel—chromium alloy. In unstressed samples there are almost no pores, which is explained by the small difference in the diffusional mobilities of nickel and chromium atoms.

When tensile stresses act during the later stages of creep preceding the rupture of samples and immediately afterward, considerable numbers of pores, cracks, and fissures occur essentially along the grain boundaries in the area subjected to high plastic deformation—the neck (Fig. 8).

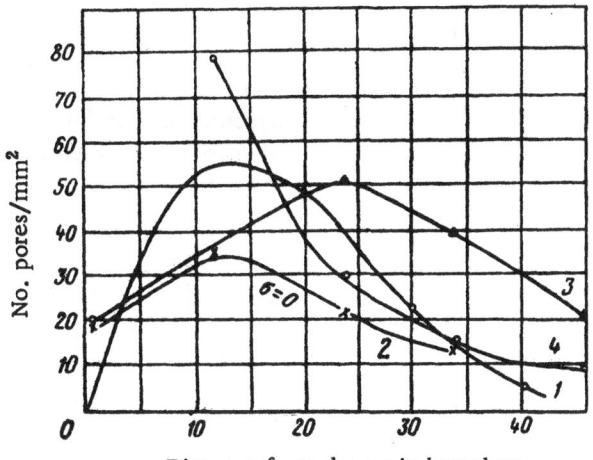

Fig. 9. Effect of stress on the pore distribution through
the grain in brass annealed 5 h at 650°C. 1) Calculated
curve; 2) unstressed sample; 3, 4) stress of 21 kg/cm².

Thus, destruction at high temperatures is also related to the process of pore formation in this case. However, the process of pore formation is not purely diffusional in the sense that the excess concentration of vacancies necessary for the formation of pores does not result from the difference in diffusional flows.

It should be noted that the formation of pores leads to considerable weakening of the grain boundaries. Thus, if pores are formed in unstressed brass samples at 750°C then subsequent testing at room temperature leads to destruction under very low tensile stresses and the fracture runs along the grain boundaries rather than through the grain.

We made measurements of the evaporation rates for brass and nichrome samples. We determined the activation energy of the process in the initial and final stages, which showed that in the first case the value of Q is almost equal to the heat of sublimation (E) of the volatile component—zinc (Q = 31,000 cal/g-atom and E = 31,200 cal/g-atom). In the second case, i.e., after prolonged evaporation, the activation energy of

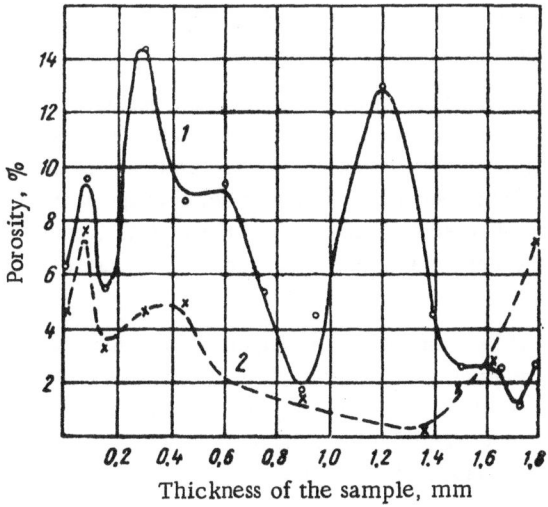

Fig. 10. Distribution of pores through a brass sample annealed at 800°C. 1) Stressed sample; 2) unstressed sample.

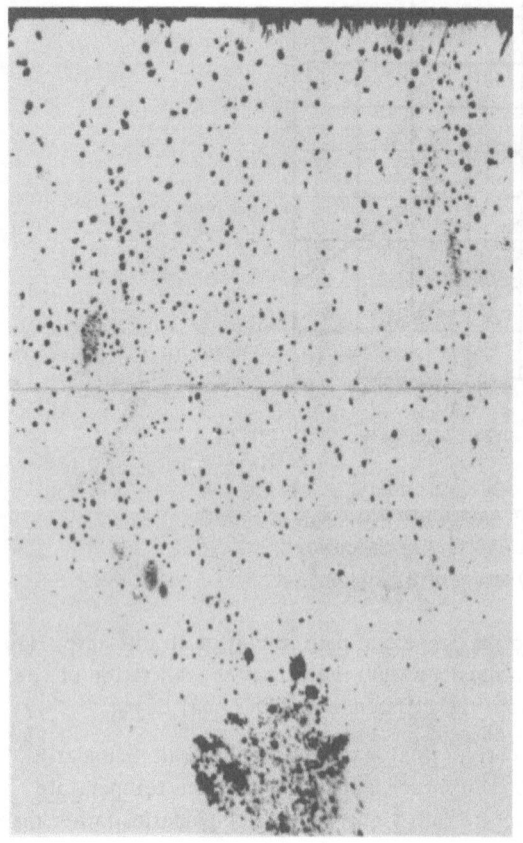

Fig. 11. Accumulation of pores in brass containing
inclusions. ×100.

the process (Q = 16,000 cal/g-atom) is close to the activation energy of the diffusion of zinc along the grain boundaries. This result indicates that diffusion becomes the controlling factor during the later stages of evaporation, when the volatile component within the metal is being evaporated.

For nichrome, the energy of evaporation remains almost the same during the whole period of isothermal annealing, and is close to the heat of sublimation of nickel and chromium (Q = 100,000 cal/g-atom, E_{Ni} = 101,000 cal/g-atom, and E_{Cr} = 95,000 cal/g-atom).

Effect of Structure on Pore Formation

We made a theoretical calculation of the distribution of pores in a grain, taking into account the diffusion rate of zinc and copper in brass. It was shown in [3] that all the additional vacancies formed during the early stages of the evaporation process move to the surface. During later stages of evaporation some of the vacancies move to the surface and others, in approximately equal numbers, remain within the sample. Therefore, the maximum number of vacancies in the metal—and consequently the maximum number of vacancies coagulated into pores—must exist at a certain distance from the surface. Figure 9 shows the experimental pore distribution curve per mm^2 of the surface in α-brass after evaporation at 600°C for 5 h under a stress of 21 kg/cm^2 and in unstressed samples. For comparison, we show the theoretical curve in the same figure. The position of the maximum on the experimental curve for the unstressed sample coincides with the position of the maximum on the theoretical curve, but the position of the maximum for the stressed sample corresponds to a greater depth than that on the theoretical curve. However, it is impossible to draw any conclusion on the distribution of pores because the results are not very reproducible, apparently due to the effect of structure and the purity of the alloy.

Figure 10 shows that the maxima are distributed somewhat periodically over the cross section of the sample. The maxima are much clearer for stressed samples. In some cases these maxima corresponds to the surface of the sample. But if such a sample is electropolished before diffusional annealing, then the pore distribution curve has a maximum corresponding to a certain depth of the sample. This distribution of pores could be due to the distribution of defects or impurities in the metal. Apparently these defects or impurities do not have time to move to the surface and form pores close to the surface under conditions in which the coagulation of vacancies is rapid.

Theoretical analyses made by a number of investigators show that the formation of the nucleus of a pore of critical size in pure metals requires a high degree of supersaturation. At the same time, the presence of ready-made separation surfaces decreases the necessary supersaturation considerably. Such nonuniform distribution and accumulations of pores resulting from the heterogeneity of the material, particularly that resulting from the presence of inclusions, is shown in Fig. 11.

Sometimes the pores are regularly distributed in lines varying from a few hundredths of a mm to 2 mm long. This distribution is strictly periodic, probably because of the periodicity of the crystal lattice. Possibly the formation of pores is facilitated in areas of accumulations of dislocations on wide-angle boundaries (Fig. 12).

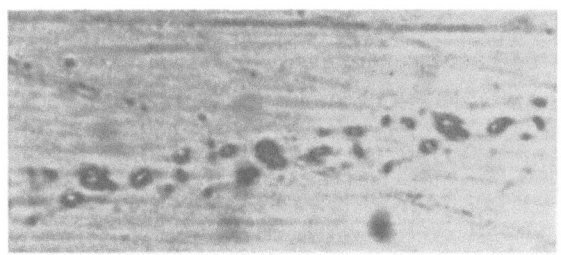

Fig. 12. Distribution of pores in a brass sample after
annealing. ×200.

Grain boundaries change the nucleation and distribution of pores. During evaporation, the volatile component first leaves the grain boundaries and the areas adjacent to the boundaries. The excess vacancies necessary for the formation of pores are formed first in boundary areas.

Impurity atoms can be absorbed at the grain boundaries, and thus create additional deformation of the lattice, thus facilitating the process of pore formation.

On the other hand, the grain boundaries are absorbers of vacancies, and as the result the supersaturation of vacancies in boundary areas must decrease. The ratio between these factors controls the process of pore formation at the grain boundaries.

To determine the mechanism of the preferential transfer of the evaporating component, we used the Fisher model, which is acceptable in our case, since our samples had large grains, and the mutual effect of boundaries is apparently small.

We determined the weight loss (%) in samples of α-brass as a function of the annealing time. The measurements showed that the volatile component reaches the surface of the sample by boundary diffusion.

Thus, we have shown that the process of evaporation and the accompanying diffusional redistribution of components of α-brass are accompanied by the formation of pores whose rate of formation at a given temperature depends on the applied stress and on the structure of the metal (grain boundaries).

Under the effect of stress and deformation the intensity of pore formation increases, and the role of the grain boundaries in diffusion also increases.

The process of pore formation terminates in the formation of cracks and destruction along the grain boundaries.

When the diffusion rates of the components of the alloy (nickel and chromium in nichrome) are about the same, almost no pores are formed. However, when tensile stress is applied the areas of this alloy subjected to great plastic deformation show the existence of considerable numbers of pores, cracks, and the fracture runs essentially along the grain boundaries.

We have shown experimentally that the nonuniformity of the material has a considerable effect on the distribution of pores.

Studies of the process of pore formation under isothermal conditions and the dependence of the process on the temperature and stress make it possible to investigate the destruction of metals at high temperatures, where the elementary processes are diffusion and pore formation and the determining structural factor is the grain boundary.

LITERATURE CITED

1. R. W. Balluffi and B. H. Alexander, J. Appl. Phys. 23 (9): 953, 1952; R. W. Balluffii and L. L. Seigle, Acta Met. 3 (2): 170, 1955; R. W. Balluffii and L. L. Seigle, J. Appl. Phys. 25 (5): 607, 1954; Ya. A. Geguzin, Usp. Fiz. Nauk 11 (2): 217, 1957; Macroscopic Defects in Metals, Metallurgizdat, 1962.
2. S. Z. Bokshtein, T. I. Gudkova, A. A. Zhukhovitskii, and S. T. Kishkin, Some Problems of the Strength of Solids, Izd. Akad. Nauk SSSR, Vol. 76, 1959.
3. Ya. E. Geguzin and H. H. Ovcharenko, Fiz. Metal. i Metalloved. 4 (3): 400, 1957.

EFFECT OF DEFORMATION ON PORE FORMATION

S. Z. Bokshtein, T. I. Gudkova, A. A. Zhukhovitskii, and S. T. Kishkin

The formation of pores during degassing is due essentially to the difference in the diffusion rates of the components of the alloy.

We investigated the effect of plastic deformation on pore formation particularly because it was shown earlier that plastic deformation increases the diffusion rate of tin in nickel along the boundaries and also through the grains.

We used samples of α-brass containing 38% zinc and 62% copper. Cylindrical samples 10 mm long were cut out of rods 15 mm in diameter. These samples were annealed in an argon atmosphere at 800°C for 3 h.

The annealed samples were subjected to different degrees of deformation in a press at room temperature (3-5, 15-20, and 55-60% deformation).

Four samples were placed in the furnace together, three of them deformed to different degrees and the fourth did not deform.

The samples were subjected to diffusional annealing in a vacuum furnace at 700, 800, and 900°C. The weight loss of the samples during annealing was taken as the characteristic of the effect of deformation on evaporation.

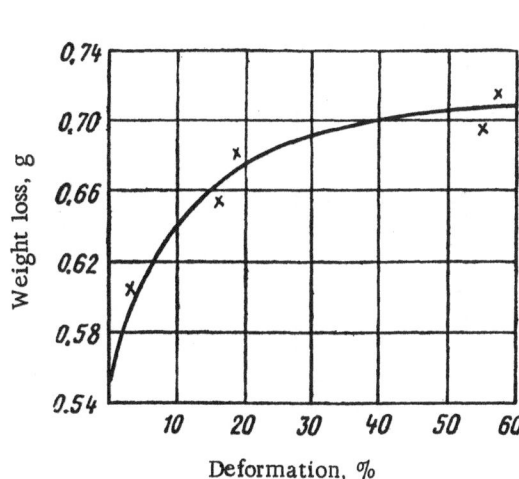

Fig. 1. Effect of plastic deformation on the weight loss of brass during annealing at 700°C for 4 h.

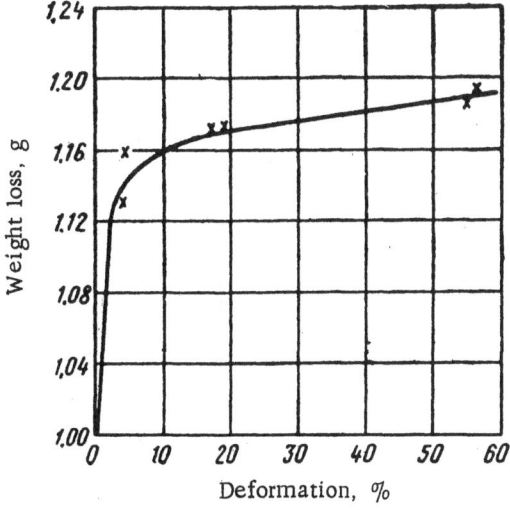

Fig. 2. Effect of plastic deformation on the weight loss of brass during annealing at 800°C for 4 h.

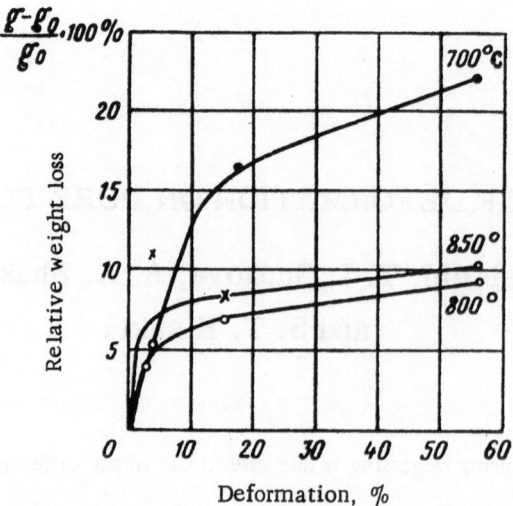

Fig. 3. Effect of plastic deformation on the relative weight
loss of brass during diffusional annealing for 4 h.

Figure 1 shows the weight loss of samples during diffusional annealing as a function of the degree of deformation. The initial section of the curve (relative to low degrees of deformation) shows a considerable increase in the rate of evaporation, but with increasing degrees of deformation the curve flattens out, indicating that the rate of increase of evaporation decreases with further increase in the degree of deformation.

The shape of the curve changes somewhat (Fig. 2) when the diffusional annealing temperature is increased to 800°C. The weight loss of the samples increases sharply when the degree of deformation is low (3-5%); the rate of increase in evaporation decreases with further increases in the degree of deformation.

At an annealing temperature of 850°C the effect of deformation is lower than it is at lower annealing temperatures. It should be noted that at this temperature the data (weight loss) are scattered.

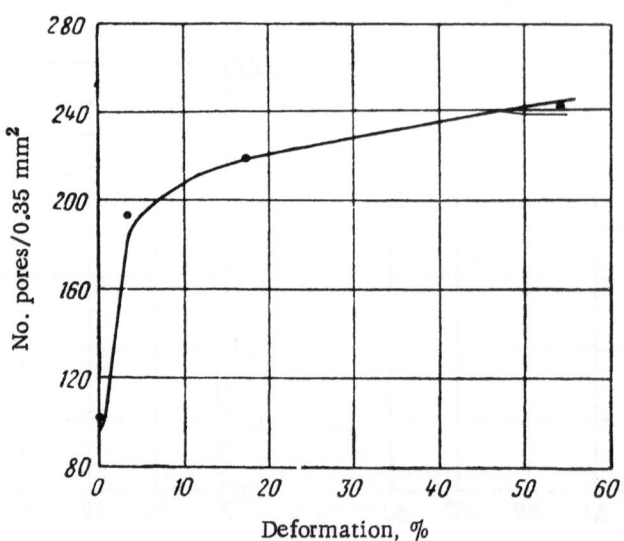

Fig. 4. Effect of plastic deformation on the number of pores
formed in brass during annealing at 800°C for 4 h.

Figure 3 shows the variation of the relative weight loss of the samples during diffusional annealing as a function of the degree of deformation. It can be seen that the effect of deformation on evaporation is greatest at 700°C. At higher temperatures (800 and 850°C) the curves practically coincide, while the effect of deformation on the relative weight loss of the samples is high only when the degree of deformation is low.

Metallographic analysis of the samples after diffusional annealing at different temperatures showed that in all cases the rate of coagulation of pores increases with the degree of deformation. The higher the temperature of diffusional annealing and the greater the degree of deformation the larger the size of the pores.

The variation of the number of pores at the surface of the sample in an area of 0.35 mm^2 with the degree of deformation is shown in Fig. 4.

Thus, plastic deformation increases the evaporation rate and also the rate of pore formation. The number and the size of pores formed during diffusional annealing increase with the degree of plastic deformation and with the annealing temperature. The acceleration of the pore formation rate and the evaporation rate under the effect of plastic deformation is due essentially to the acceleration of the diffusional mobility of atoms in the deformed material.

The change in the mobility of atoms as the result of deformation is due directly to the change in the fine structure of the metal which results from residual deformation.

In deformed samples the pores should form in the area with high concentrations of vacancies. Since plastic deformation changes the state of the grain, the pores may form either along the grain boundaries or within the grain.

The fact that the effect of plastic deformation on the evaporation rate decreases at high temperature is due to the partial healing of defects induced by plastic deformation and recrystallization.

The acceleration of the evaporation rate as the result of low degrees of deformation as compared with high degrees of deformation is apparently due to the fact that the main effects resulting from the formation of blocks and from the increase in the dislocation density are stronger during the initial stages of deformation. Further increase in the degree of deformation changes the fine structure of the metal very little.

Thus, we have shown that plastic deformation increases the evaporation rate. The number and size of pores formed at high temperatures also increase as the result of plastic deformation.

The highest increases in the evaporation rate and in the pore formation rate correspond to relatively low degrees of deformation (up to 10%); higher degrees of deformation induce very little further acceleration of the evaporation rate or the rate of pore formation. The effect of plastic deformation decreases with increasing temperature.

EFFECT OF PORE FORMATION ON DIFFUSION

T. I. Gudkova

Since displacement of atoms by diffusion plays an important role in pore formation, it is of interest to investigate the effect of the presence of pores on the diffusional mobility of the components of the alloy.

We investigated α-brass composed of 32% zinc and 68% copper. Samples $10 \times 10 \times 20$ were annealed in an argon atmosphere at 800°C for 3 h to relieve stresses. The samples were heated at 650 and 700°C for 30-40 h to induce pores. One side of the samples was coated electrolytically with a thin layer of copper so that zinc could evaporate from only one side of the sample. The thin layers used for autoradiographic analysis were cut at an angle to the surface.

A thin layer of Ni^{63} was deposited electrolytically on the surface of samples from which zinc evaporated. One side of another sample of α-brass without pores was also coated with radioactive nickel.

Porous and nonporous samples of brass were subjected to diffusional annealing at 650, 700, 800, and 850°C for different periods of time.

The diffusion constants of nickel in α-brass were determined by the absorption method and the distribution of nickel in the samples was determined from the autoradiograms of samples cut at an angle to the surface coated with radioactive material.

Under the conditions used, the presence of pores in the samples has relatively little effect on the diffusional mobility of nickel in α-brass. The diffusion coefficient in porous brass is 1.5-2 times higher than in nonporous brass.

The activation energy of the diffusion of nickel in nonporous brass is 39,000 cal/g-atom. In porous samples the activation energy decreases only 2,000 cal/g-atom.

The autoradiograms of the diffusion of nickel in α-brass show that the diffusion is general at the surface of the sample and just below the surface. At greater depths diffusion is preferential along the grain boundaries.

If we assume that the rate of diffusion through the pores is greater than in the rest of the material, then the radioactive isotope should accumulate at a certain depth corresponding to the depth of the pores. However, the analysis of the autoradiogram did not show any such accumulation in any area of the diffusion zone.

Microstructural analysis of the sample after preliminary heating showed that the number and size of pores remain practically constant during diffusional annealing.

These results can be explained by the fact that the large pores cannot play any significant role in the diffusion process. Since the excess vacancies formed as the result of the difference in the diffusion rates of copper and zinc in α-brass have time to coagulate and form pores, an equilibrium concentration of vacancies is established after preliminary heating under the conditions of the experiment. This leads to a relatively small difference in the diffusion coefficients of porous and nonporous samples.

We may also assume that the presence of microscopic pores in α-brass (which we could not detect with the magnification used in our experiments) affects the diffusion of atoms during subsequent heating of the metal. This assumption is confirmed by the fact that the diffusion coefficient of nickel in porous brass is about twice that in nonporous brass and that the activation energy of the diffusion process decreases from 39,000 cal/g-atom for nonporous samples to 37,000 cal/g-atom for porous samples.

Thus, large pores do not increase the diffusional mobility. Possibly the discontinuities in the material constitute a barrier to diffusional flow. On the other hand, small pores probably increase the diffusion rate.

PART IV

THREAD-LIKE CRYSTALS AND HIGH STRENGTH

MECHANICAL PROPERTIES OF COPPER, NICKEL, AND COBALT WHISKERS

S. Z. Bokshtein, S. T. Kishkin, and I. L. Svetlov

Thread-like crystals (or whiskers) were the first materials in which the theoretical strength of the material was attained. For example, the ultimate strength of individual whiskers of graphite, sapphire (α-Al_2O_3), and iron under tension is 2000, 1500, and 1300 kg/mm^2, respectively, while in ordinary crystals it is 20-40 kg/mm^2.

It has been known for a long time that in practice the strength of solids is one-tenth to one-hundredth the theoretical value calculated on the basis of the force of mutual attraction of atoms in an ideal crystal lattice.

In practice, the strength of materials depends not so much on the character of atomic interactions as on the defects in the crystal structure. Superhigh strength has been obtained in whiskers of pure metals, semiconductors, ionic crystals, and molecular compounds. Apparently, the crystal lattice in whiskers contains few or none of the defects responsible for plastic deformation.

The object of the present investigation was to obtain whiskers of copper, nickel, and cobalt and determine their ultimate strength. We studied the effect of alloying on the mechanical properties of whiskers and attempted to put in evidence dislocations in copper microcrystals by selective etching.

EXPERIMENTAL METHOD

Copper, nickel, and cobalt whiskers were grown by reducing anhydrous halide compounds (marked "pure") of the respective metals with hydrogen [1].

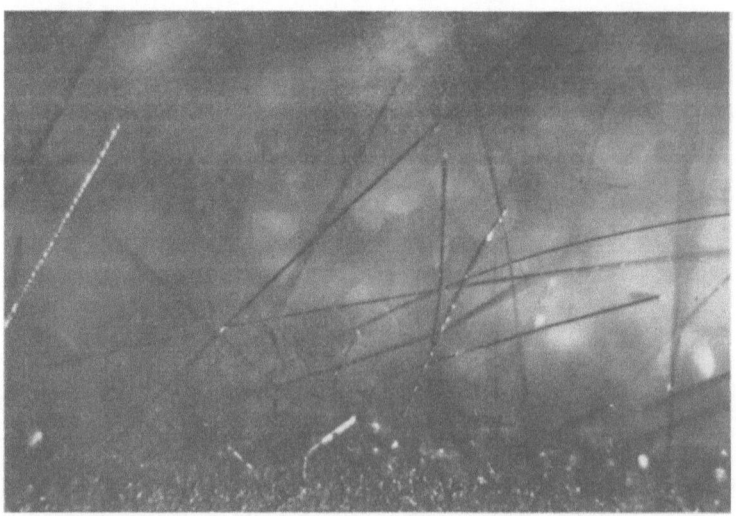

Fig. 1. General view of thread-like crystals of nickel obtained by reduction of nickel bromide. ×12.

Optima Parameters for Growing Copper, Nickel, and Cobalt Whiskers

Material	Salt	Temp., °C	Rate of hydrogen flow, cm^3/min	Time, h	Maximum whisker length, mm
Copper	CuI	580-590	0.07-0.11	1	20
Nickel	$NiBr_2$	700-710	0.02-0.04	1	7
Cobalt	$CoBr_2$	760-780	0.15-0.2	0.5	15

The apparatus consisted of a quartz reaction tube heated in a tubular resistance furnace, a source of hydrogen (electrolyzer), and a gas purifier which also regulated the gas flow in the apparatus. The hydrogen was thoroughly dried and deoxygenated by passing it through incandescent titanium sponge.

Thread-like crystals (Fig. 1) grow on the bottom and on the walls of the porcelain combustion boat in which the salt is being reduced. The whiskers differ in length (from 1 to 20 mm), in thickness (from 1 to 50 μ), and shape. There were thin straight whiskers with a smooth surface, deformed whiskers, whiskers in the shape of needles, ribbons, spirals, etc. The growth of whiskers is considerably influenced by random factors which cannot always be accounted for. However, there are optimum parameters (see the table) at which the whiskers acquire the desired dimensions and shapes.

The surface of cobalt whiskers is never absolutely smooth because of the phase transformation (Crystal 12 → Gas 12). The polymorphic transformation of face-centered cubic cobalt into hexagonal crystals at 417°C is accompanied by a small change in volume which, in spite of its smallness (0.3%) induces the formation of cobalt whiskers with shapes resembling slip planes. It is quite interesting that later on these Chernov-Lüders bands do not affect the mechanical properties of the whiskers, although it is well known that when they occur during plastic deformation they decrease the strength of the same cobalt crystals by a factor of 10-100.

Some thread-like cobalt crystals contain a network of surface planes twisted into a spiral, although an x-ray investigation of such whiskers made recently [2] did not show any twisting of the lattice. In fact, the presence of an axial screw dislocation in whiskers must be accompanied by a twisting of the crystal lattice. However, it is not clear why such twisting is visible with an optical microscope but is not revealed by x-rays.

The whiskers must be removed from the combustion boat and attached to the holder in the testing apparatus with great care. We developed a special method of handling whiskers.

1. The "good" whiskers are selected with a binocular microscope with a magnification of 10-20.

2. The whiskers are cut off at the base with an electric arc and removed from the boat.

3. The samples are glued to the holder of the testing apparatus.

4. The side faces of the whiskers are examined metallographically with the MBI-6 apparatus under a magnification of 500-600.

All the operations concerning the manipulation and mounting of the whiskers are carried out with a Zeiss micromanipulator and other instruments such as single and double probes, microsolderer, etc.

The whiskers were placed under tension in the apparatus mounted on the base of the ADV-200 analytical balance.

Samples 1-15 μ in diameter and 1-10 mm long can be tested at room temperature under loads of 0.5-7 g in this apparatus, which consists of three main parts: a loading and force measuring mechanism consisting of a solenoid with constant current 11, with an iron core 10, suspended from one shoulder of the balance, a horizontal microscope 4, and a mechanism for coaxial gluing of the samples. The sample is glued in the clamps of the microapparatus with diphenylcarbazide glue. By increasing the current in the coil of the solenoid one can gradually load the sample until it is ruptured. The maximum stress withstood by the crystal before destruction is considered to be the ultimate strength in tension.

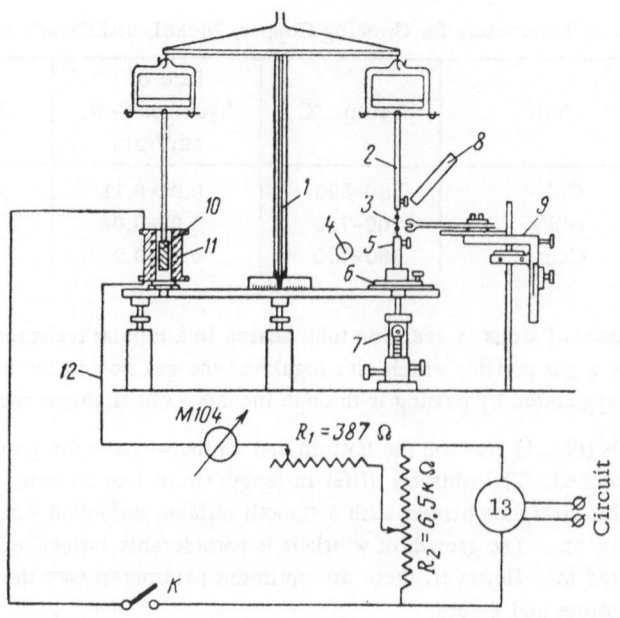

Fig. 2. Diagram of the microapparatus for tensile strength tests of whisklers. 1) ADV-200 analytical balance; 2) upper holder; 3) whiskers; 4) horizontal microscope; 5) lower clamp; 6) ST-12 sample mover; 7) vertical lift mechanism; 8) OI-19 lamp; 9) microsolderer; 10) iron core; 11) solenoid; 12) electric furnace, 13) stabilizer.

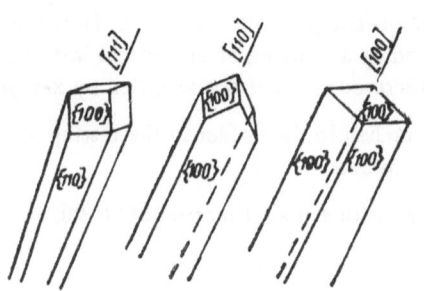

Fig. 3. Relationship between the shape of the cross section of the whisker and its orientation.

To determine the ultimate strength of whiskers it is necessary to determine the cross section with the greatest precision. To determine the cross sections of whiskers from linear dimensions measured under a microscope may lead to considerable error, particularly when the diameters are so small. For example, if one considers the hexagonal or square cross section as a circle with a diameter equal to that of a circle inscribing the hexagon or square the error in determining the cross section is 17 and 30%, respectively. In the case of cross sections with more complex shapes such calculations lead to an error exceeding 50%.

The method of determining the shape and dimensions of the cross sections of the whiskers consisted in cutting cross sectional samples from the whiskers [3]. The cross sectional samples were photographed under a magnification of 2000 and the cross section was determined from the photographs with a planimeter. The square root of the area of the cross section was taken as the diameter of the whisker.

It is well known that one can determine the crystallographic orientation of the longitudinal axis of the crystal from the shape of the cross section of the whisker. The hexagonal cross section indicates the [111] orientation, a square indicates the [100] orientation, and a rectangle the [110] orientation (Fig. 3).

The preferential orientations of copper and cobalt whiskers tested in this work were close to [111] and [0001], while that of nickel whiskers was close to [100] or [110] as indicated by the shape of the cross section (Fig. 4). Quite often the cross sections of the whiskers have an irregular shape. Sometimes a square cross section terminates in a hexagonal cross section, and therefore the precise determination of the strength of the whiskers is a rather complicated problem.

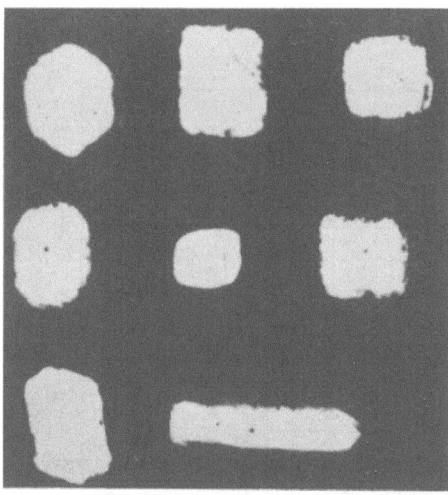

Fig. 4. Shapes of the cross sections of whiskers of copper, nickel, and cobalt. ×2000.

RESULTS

Using the method described, we placed 35 copper whiskers, 34 nickel whiskers, and 20 cobalt whiskers under tension. The length of the samples varied from 1.5 to 3 mm and the diameter from 2 to 15 μ. The thinnest whiskers (2-3 μ) were the strongest (the maximum strength of whiskers is 8-10 times higher than the strength of ordinary massive single crystals). The effect of the diameter on the ultimate strength of copper, nickel, and cobalt whiskers is shown in Fig. 5. In spite of the considerable scattering of the results, it can be seen that the strength decreases with increasing diameter of the whisker. The same relationship has been found by other investigators for copper and iron [4], and silicon [5], but was shown here for nickel and cobalt for the first time.

The highest point on the curve corresponds to the theoretical strength which, according to different authors, is equal to $G/6 - G/30$, where G is the shear modulus. The lower end of the curve (showing the strength of whiskers with the largest diameters) approaches the ordinary strength of massive single crystals (15-30 kg/mm^2). It should be noted that a sharp increase in the strength of whiskers occurs only for whiskers with small diameters (2-3 μ). The dependence of the strength on the diameter can be expressed as follows:

$$\text{For copper} \quad \sigma_B = \frac{308}{d} + 16.8, \tag{1}$$

$$\text{For nickel} \quad \sigma_B = \frac{620}{d} + 23, \tag{2}$$

$$\text{For cobalt} \quad \sigma_B = \frac{1200}{d} - 45. \tag{3}$$

Equation (3) is valid for diameters less than 25 μ. Let us note that this dependence of the ultimate strength on the diameter is also found in the case of thin glass threads.

As we have said, the whiskers have different orientations. Investigations of the strength of whiskers with known orientations showed that, within the experimental error, the difference in orientations does not affect the strength, as can be seen in Figs. 5, 6, and 7.

This fact contradicts the conclusions in [6], where it was found that the strength of whiskers depends on the orientation.

We also investigated the variation of the strength along the length of copper whiskers. Copper whiskers 3 μ in diameter and 8 mm long were cut in two with an electric arc. Each half was tested separately. One half had an ultimate strength of 163 kg/mm^2 and the other 10 kg/mm^2. This experiment indicates the statistical distribution of defects along the whisker.

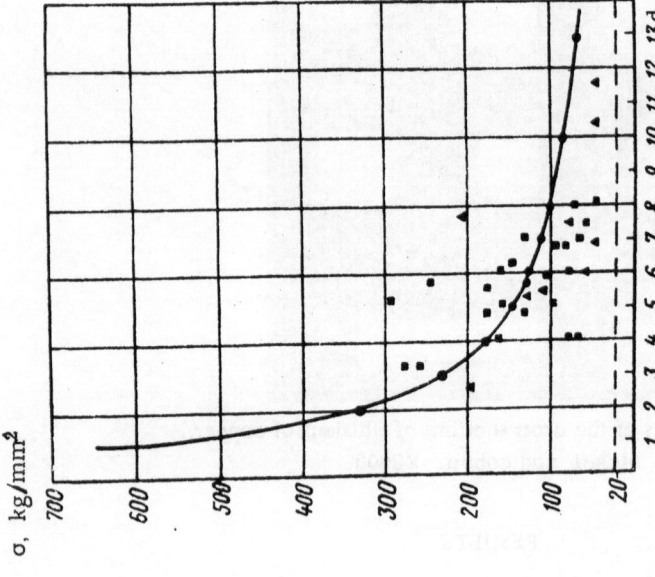

Fig. 6. Effect of the diameter on the ultimate strength of nickel whiskers under tension. ▲) [111] orientation; ■) [100] or [110] orientation. The points on the curve were calculated by the method of least squares.

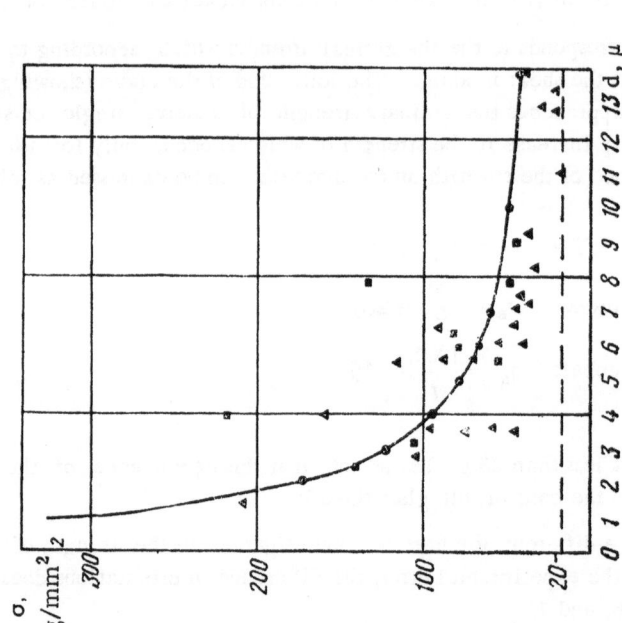

Fig. 5. Effect of the diameter on the ultimate strength of copper whiskers under tension. ▲) [111] orientation; ■) [100] or [110] orientation. The points on the curve were calculated by the method of least squares.

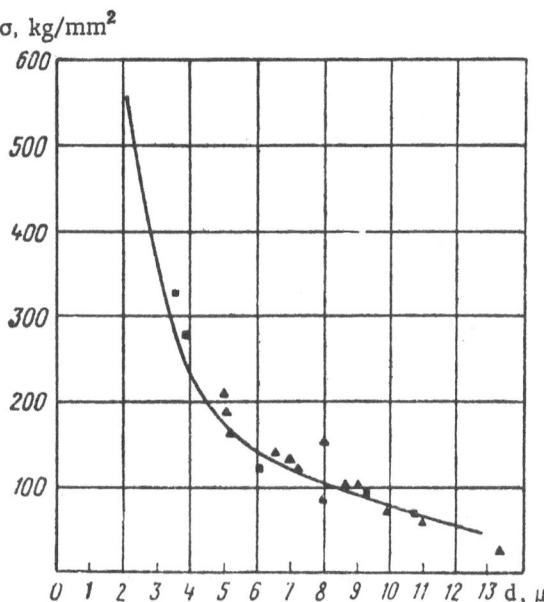

Fig. 7. Effect of the diameter on the ultimate strength of cobalt whiskers under tension. The notations are the same as in Fig. 6.

The surface state of the whiskers has a great effect on the strength. Often one finds such defects on the surface of whiskers as protrusions, branched threads, pits resulting from pitting corrosion, and sometimes grain boundaries. The strength of whiskers containing such surface defects is very low. Preliminary experiments on the effect of oxide films and a deposited film of silver on the surfaces of copper whiskers showed that such films do not induce any significant changes in the strength, although in some cases a thick oxide film on copper whiskers produced in an ammonia atmosphere decreases the strength.

All the whiskers investigated underwent considerable plastic deformation before rupture, particularly cobalt, in which the plastic flow sometimes reached 500%. Very thin nickel whiskers twisted into spirals after rupturing. The observation of the process of plastic deformation under the microscope showed how slip lines propagate (Fig. 8). Very often the slip occurs in intersecting planes. During plastic deformation of copper a neck is formed, although the rupture does not always occur in the neck, which apparently indicates that this portion of the sample becomes stronger. Although the strength of whiskers usually drops drastically at the beginning of plastic deformation, in some cases a reverse effect was found: strengthening following deformation. Possibly this occurs in whiskers containing a great number of defects. It is well known that in ordinary crystals several consecutive necks sometimes form. Knowledge of the effect of alloying on the mechanical properties

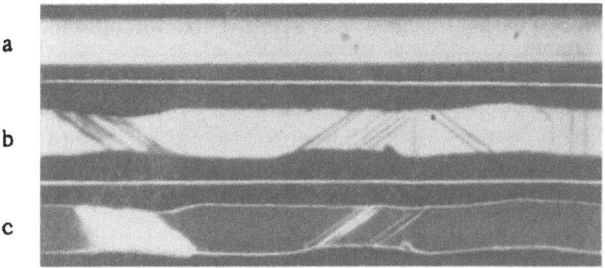

Fig. 8. Development of slip planes in copper whiskers. ×800. a) Crystals before deformation; b, c) crystals after rupture, in dark and light fields.

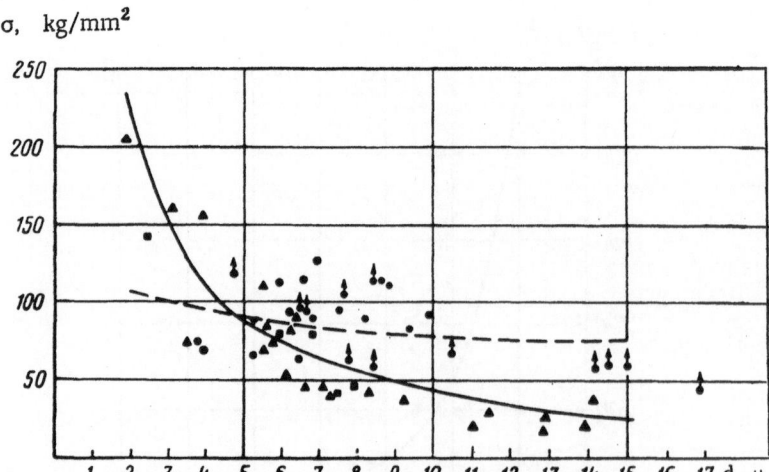

σ, kg/mm²

Fig. 9. Effect of alloying on the strength of copper whiskers. The solid
line corresponds to pure copper and the dashed line to the copper—silver
alloy. ●) Alloy whiskers; ■, ▲) pure copper whiskers.

is very important for understanding the causes of the high strength of whiskers. For this purpose we investigated
the effect of silver on the strength of copper whiskers. The whiskers were alloyed by diffusional saturation.
Silver was deposited in vacuum (10^{-4} mm Hg) on copper whiskers and then the whiskers were annealed in a
stream of hydrogen at 800°C for 1-2 h.

The annealing conditions were chosen so that the silver diffused through the whole cross section of the
thickest whisker. The results of tensile strength tests of Cu—Ag whiskers are shown in Fig. 9. Let us note the
following singularities in these results.

When the diameter of the whiskers is large the curve for the alloy lies above the curve for the pure metal.
If we assume that the decrease of strength in whiskers with relatively large diameters is due to defects in the
crystal lattice, then the observed increase in strength resulting from alloying copper whiskers with silver can
be explained in first approximation by the ordinary mechanisms of strengthening of defective crystals. These
mechanisms can be related to the formation of solid solutions or a second phase or the capacity of aging. It is
well known that Cu—Ag alloys become stronger as the result of dispersional aging.

When the diameter of the whiskers is small the strength of the alloy whiskers is lower than the strength
of pure metal whiskers, and this is due to the fact that during the formation of a second phase a large number
of sources of dislocations may be formed, which leads to a decrease in strength.

Unlike whiskers of pure metals, the whiskers of alloys always have circular cross sections (see Fig. 4).
This is due to the fact that the deposited film of silver flows as the result of surface diffusion during annealing
and tends to acquire the minimum surface energy.

To understand the mechanism of plastic deformation occurring in whiskers, one must compare the properties of separate dislocations or groups of dislocations with the macroscopic mechanical characteristics—the
yield strength of single crystals, for example.

In this respect, the etching method seems very promising. This method makes it possible to find not
only the distribution of single dislocations after different treatments but also to observe the displacement of
dislocations. This method has been successfully used for ionic compounds, but its application to metals is difficult because of the selective etchability of the crystallographic planes.

The etching agents recently developed for metals make it possible to put in evidence decorated dislocations in freshly grown crystals and also pure dislocations induced by plastic deformation.

The formation of etch pits at intersections of dislocation lines with the surface of the crystal is due to
the high etchability resulting from local increase in the chemical potential due to the accumulation of impurities on dislocations or to the elastic stress fields.

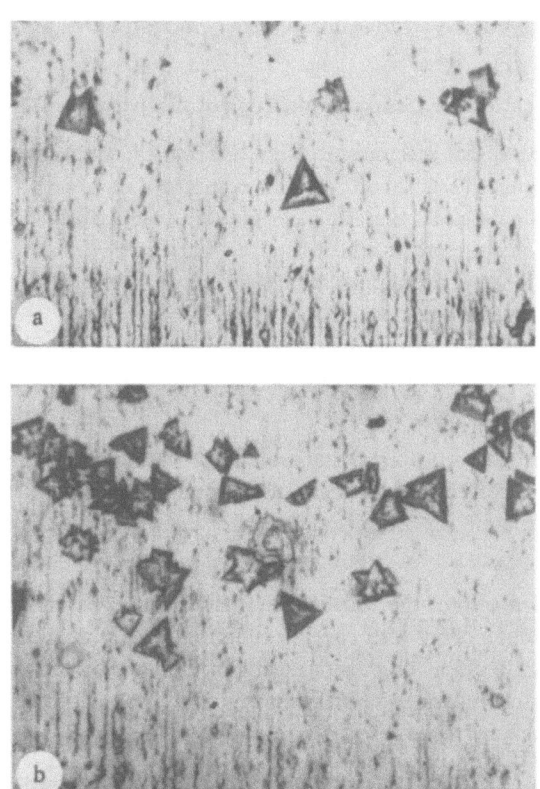

Fig. 10. Dislocations in the (100) plane of copper microplates. ×420.

Fig. 11. Block boundaries with "bamboo" structure in (100) plane of copper microcrystals. ×420.

119

Copper microplates grown by the method used for whiskers but at higher temperatures were etched with Jouhg's reagent [7], which consists of a 2M solution of $FeCl_3 \cdot 6H_2O + 7.8M$ HBr.

The samples were etched in this electrolyte for 10 sec. Since these experiments have only begun, we shall give only a few results.

First of all, there are dislocations in freshly grown copper microcrystals (Fig. 10).

Secondly, it was demonstrated that microcrystals are not single crystals but consist of separate crystals oriented at small angles to each other (1-2°), but having a common axis of growth. Figure 11 shows such a small-angle boundary between two crystals consisting of separate dislocations. Using the x-ray method, Dragsdorf recently found the same block structure in whiskers after martensitic transformation. Finally, the dislocation density is very high at the places where the crystals were cut from the base. This work on the dislocational structure of copper microcrystals is being continued.

DISCUSSION OF RESULTS

The scattering of results on the variation of the strength of whiskers as a function of the whiskers' diameters indicates that the strength of perfect crystals decreases as the result of statistically distributed defects. These defects induce premature rupture of whiskers either after considerable plastic deformation or without deformation.

Possibly, the change in the length of the whiskers is an additional reason for the scattered results. Not only the statistical nature of the defects but also careless handling during mounting of the whiskers in the clamps of the apparatus have a considerable effect on the scattering of the data on the properties of whiskers. There are data [8] indicating that in places where the whiskers are glued there is sometimes formed a network of dislocations, which undoubtedly decreases the strength.

Apparently, the defects which decrease the strength and lead to irregularities in the strength are introduced into the crystal during growth, and therefore the density of defects depends first of all on the growing conditions. By changing such a factor as the degree of purity of the initial material or the reducing gas one can obtain crystals almost free from defects.

The scattering of the results on the strength would be even greater if the whiskers were not carefully chosen before the test. We would say that on the average there is only one good whisker per hundred.

The defects which decrease the strength of whiskers are localized on the surface and also within the crystal. The gross surface defects can be seen with a microscope. Samples with such defects were not tested. But surface defects which cannot be seen with an optical microscope also decreases the strength of the whiskers.

Among the internal defects which decrease the strength of whiskers are point defects of the vacancy type, penetration impurity atoms, and also linear defects of the dislocation type.

According to the theory of growth, the high-strength whiskers have a screw dislocation along the axis of the whisker; sometimes one finds two, three, and even more dislocations of this type. The crystals having a high dislocation density have a low strength. But the problem is then to determine the degree of "safe" dislocation density (which would not decrease the strength of whiskers) and also to determine the distribution of dislocations. Apparently it can be assumed that the high strength of thread-like crystals is due to their perfect structure, and not the scale factor, although this also has a definite effect.

LITERATURE CITED

1. S. S. Brenner, Acta Met. 4: 62, 1956; Hauptman, Ya. Katser, and R. Gemperle, Crystal Growth, Coll. edited by Shubnikov, Izd. Akad. Nauk SSSR, 1961.
2. D. Dragsdorf and Johnson, J. Appl. Phys. 33 (2): 724, 1962.

3. S. Z. Bokshtein and I. L. Svetlov, Zavodskaya Lab. (5):595, 1962.

4. S. S. Brenner, Growth and Perfections in Crystals, 1958; I. P. Kushnir and Yu. A. Osip'yan, Dislocations in Metals and Problems of Strength, Izd. IMET, Moscow, 1961, p. 11; I. A. Oding and I. M. Kop'ev, Metalloved. i Term. Obrabotka Metal., No. 9, 1961.

5. E. M. Nadgornyi and A. V. Stepanov, Fiz. Tverd. Tela, Akad. Nauk SSSR 3 (4):2068-2073, 1961.

6. I. A. Oding and I. M. Kop'ev, Metalloved. i Term. Obrabotka Metal., No. 9, 1961.

7. F. W. Jouhg, J. Appl. Phys. 32(2):192, 1961.

8. P. B. Price, Phil. Mag. 5 (52):417, 1960.

GROWING SAPPHIRE WHISKERS

S. Z. Bokshtein, M. P. Nazarova, and I. L. Svetlov

Aluminum oxide whiskers (sapphire)— α -Al_2O_3—are presently the most promising thread-like crystals for practical applications. Thus, there is great interest in the problem of growing sapphire whiskers. Several variations of essentially the same method have been proposed. This method consists of high-temperature oxidation of aluminum in an atmosphere of humid hydrogen.

The initial material is either a powder of pure aluminum or aluminum oxide. The crystals are grown in an atmosphere of hydrogen and water vapor (with a partial pressure of 10^{-3}-10^{-4} atm) under a pressure of 1 atm.

Thread-like sapphire crystals are the only ones for which the dislocational growth mechanism has been proved. The presence of a screw dislocation along the axis of the sapphire whiskers has been shown to exist by x-ray analysis and optical investigation. Recently, it was shown quite convincingly [1] that sapphire whiskers inherit this screw dislocation of the monocrystalline substrates on which they are grown. The maximum statistical average ultimate strength of sapphire whiskers is 1500 kg/mm^2, the average strength at 20°C being 750 kg/mm^2, while the momentary strength at 2000°C is 60 kg/mm^2 [2].

The object of this investigation was to develop a method of growing aluminum oxide whiskers and also a metallographic study of the shape and dimensions of the crystals obtained.

APPARATUS AND METHOD OF GROWING WHISKERS

The apparatus for growing aluminum oxide whiskers (Fig. 1) has three parts: a tubular furnace, P, in which six Silit rods, P-3, are heated simultaneously; a hydrogen source, E; and a system to purify and measure the gas, O.

The working chamber is a P-2 alundum tube 650 mm long with an internal diameter of 50 mm. The ends of the tube are hermetically closed with removable steel flanges and gaskets P-1. Two alundum jackets are mounted in one flange through which pass the Pt—PtRh thermocouples and a tube for the introduction of hydrogen. The second flange has a tube for the introduction of the gas and a molybdenum glass window for observation of the growth of whiskers during the experiment. The furnace is heated with a 220-V alternating current which passes through an RNO-250-10 voltage regulator. The temperature is regulated with an EPD-12 electronic potentiometer. The source of hydrogen is a battery of two electrolyzers E whose construction is described in Zavodskaya Laboratoriya, No. 2: 165, 1934.

The electrolyzers are fed with a 30-A direct current with a voltage of 4-6 V supplied through two SV-24-64 selenium rectifiers in parallel (the diagram shows only one). The yield of the electrolyzers is as high as 30 liter/h.

The hydrogen is thoroughly purified before entering the working chamber. The balloon O-10, contains glass cotton, where the large drops of alkali are removed. The hydrogen is then dried over silica gel in the U-shaped tube O-11, and finally passes into the tube containing an incandescent copper bar O-12, where oxygen is removed. After a second passage over silica gel the hydrogen contains the necessary amount of water vapor with a partial pressure of 10^{-3} mm Hg. The rate of the gas flow is measured with a capillary rheometer O-13. The usual hydrogen flow was 0.3-0.5 liter/min.

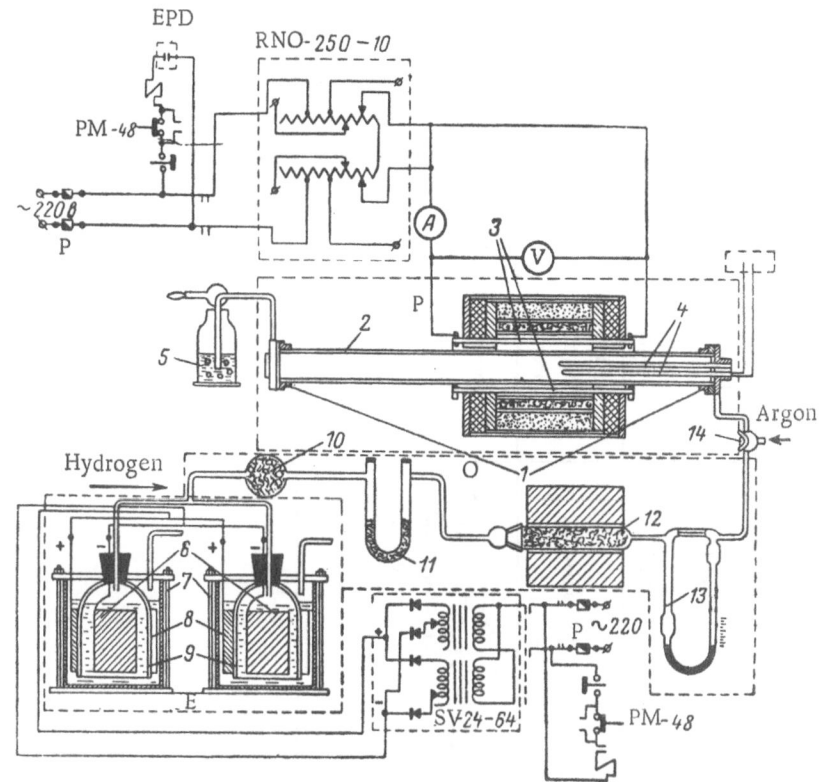

Fig. 1. Diagram of the apparatus for growing aluminum oxide whiskers. P-1)
End flange; P-2) alundum reaction tube; P-3) Silit rods; P-4) thermocouple
jackets; P-5) Drexel flask; E-6-8) Electrodes; E-7) steel support rods; E-9) bottle with bottom cut out; O-10) bottle with glass cotton; O-11) silica gel; O-12)
copper shavings; O-13) rheometer; 14) three-way valve.

The experiments with this apparatus were made in the following way.

Standard corundum combustion boats 100 and 180 mm long were first fired at 1350°C in a stream of hydrogen and then filled with the mixture of aluminum powder and 3-6% Al_2O_3 and then placed in the hot area
of the furnace (1350-1400°C).

The whole cycle of the experiment took 5-6 h (the temperature in the furnace was raised to 1400°C over
a period of 2 h); the samples were kept at this temperature 1-2 h and then cooled to 500°C in a stream of hydrogen. Before loading the furnace and before removing the samples, the furnace was flushed with an inert
gas (argon or helium) to remove the air. The different conditions under which the aluminum oxide whiskers
were grown are given in the table.

EXPERIMENTAL RESULTS

The aluminum oxide whiskers grow on the bottom and the sides of the boat. There are three distinct
zones over the length of the boat. Large crystals, which we shall call microcrystals, usually grow in the first
zone, directly against the stream of hydrogen. Very often these flat microcrystals are pointed like a sword
(Fig. 2a, 2b), or are rather thick needles (Fig. 2c) or hexagonal prisms (Fig. 2d). The end of the prism is either
widened or has a cap (Fig. 2e). Probably these caps form as the result of radial growth in the direction opposite
the initial direction. The cross section of a hexagonal prism is shown in Fig. 3.

Methods of Growing Thread-Like Crystals

Authors	Webb, Forgeng[1]	Sears, De Vries [2, 6]	Hargreaves [3]	Brenner[4]	Edwards, Happel [5]	Authors of this investigation
Temperature	1350-1450°C	1730°C	1300-1450°C	1250°C	1400°C	1360-1390°C
Mixture	Al or TiAl₃ (with a slight excess of Al)	Al₂O₃ rod	Pure Al_2O_3	Al powder	Al shavings on oxide substrates	Al powder
Composition of gaseous medium	$p_{H_2}=1$ atm $p_{H_2O}=1.2\cdot10^{-4}$ atm, dew point ≈ −70°C	Hydrogen with a dew point of ≈ −70°C	$p_{H_2}=1$ atm $p_{H_2O}=10^{-3}-10^{-4}$ atm	$p_{H_2}=1$ atm $p_{H_2O}=1.2\cdot10^{-4}$ atm, dew point ≈ −50°C	Hydrogen with a dew point of ≈ −30°C	$p_{H_2}=1$ atm $p_{H_2O}=10^{-3}$ atm
Rate of flow of hydrogen	−	1 liter/min	200-700 cm³/min	−	−	300-400 cm³/min
Growing time	4-5 h	90 min	4-5 h	4-5 h	2 h	1-2 h

1. W. W. Webb and W. D. Forgeng, J. Appl. Phys. 28 (12): 1449, 1957.
2. R. C. DeVries and G. W. Sears, J. Chem. Phys. 31 (5): 1256, 1959.
3. C. M. Hargreaves, J. Appl. Phys. 32 (5): 936, 1961.
4. S. S. Brenner, J. Appl. Phys. 33 (1): 33, 1962.
5. P. L. Edwards and R. J. Happel, J. Appl. Phys. 33 (3): 826, 1962.
6. G. W. Sears, R. C. DeVries, and C. Huffine, J. Chem. Phys. 34 (6): 2642, 1961.

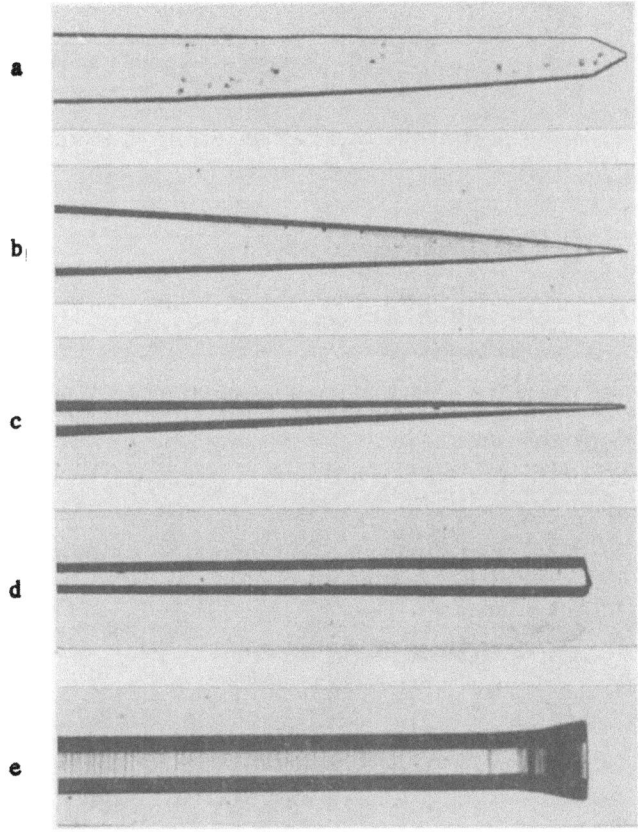

Fig. 2. Shapes of aluminum oxide microcrystals. a) Rectangular plate, ×400; b) hexagonal plate, ×180; c) needle microcrystals, ×180; d) hexagonal prism with a straight end, ×180; e) hexagonal prism with a widened end in the form of a cap, ×180. All the photographs were made in transmitted light.

Let us note that a large number of such microcrystals grow in long boats. Apparently, by simply increasing the size of the apparatus and the boats one can obtain great numbers of large aluminum oxide crystals. It is quite interesting that the hexagonal prisms and sometimes the needles grow from nuclei of the same shape formed on wide rhombohedral plates. Such a hexagonal nucleus in different stages of growth is quite visible in Fig. 4.

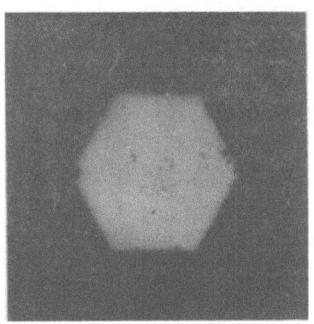

Fig. 3. Cross section of a hexagonal prism. Dark field. ×200.

Beyond the first zone is the zone of the growth of thread-like crystals itself (Fig. 5). The diameter of these crystals varies from 1 to 15 μ and the length from 10 to 15 mm. The surface is usually uniform and smooth and gives beautiful interference fringes in reflected light. However, sometimes dendrites in the form of pyramids grow on the long whiskers (Fig. 6). Quite often drops are formed at the ends of long thin whiskers. It is assumed that these are drops of pure aluminum.

The boat becomes filled with a hairy precipitate covered with a multitude of very thin and short hairs.

In the center of some plates and some prisms there is a thin capillary passing through the entire length of the crystal (Fig. 7).

Fig. 4. Hexagonal nucleus of a microcrystal. Light field. ×200.

Fig. 5. General view of aluminum oxide whiskers. ×36.

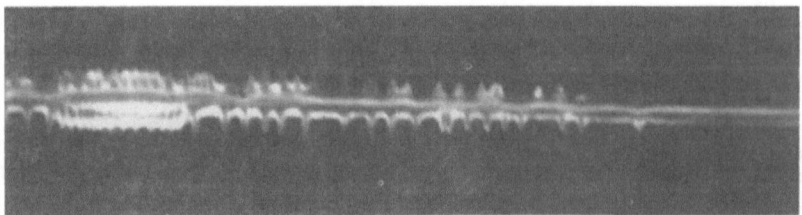

Fig. 6. Whisker with dendrite protrusions. Dark field. ×200.

Fig. 7. Hexagonal prism with a capillary running the length of the crystal. ×180. Transmitted light.

In the case of porcelain boats the crystals have an irregular surface and are quite branched. In some cases the whiskers grow from the inner walls of the reaction tube.

This fact indicates that the crystals grow in the gaseous phase.

The chemical reaction occurring during the transformation of aluminum oxide into whiskers is not yet completely clear. It is assumed that the main reaction is as follows:

$$\text{Al (liquid)} + 3H_2O \text{ (gas)} \rightleftharpoons Al_2O_3 \text{ (solid)} + 3H_2 \text{ (gas)}. \qquad (1)$$

For this reaction the ratio of the partial pressures of water vapor and hydrogen $p_{H_2O}/p_{H_2} \approx 10^{-7}$ at 1500°C. Since in our experiments this ratio was equal to 10^{-3}, reaction (1) proceeds from left to right. In spite of the fact that the vapor pressure of Al_2O_3 at 1500°C is very low, our experiments as well as data from the literature indicate that the whiskers grow from the gaseous phase. Two volatile aluminum oxides—Al_2O and AlO—are known to exist in the range of tmperatures used in our experiments. These oxides are formed according to the reactions

$$2\text{Al (liquid)} + H_2O \text{ (gas)} = Al_2O \text{ (gas)} + H_2 \text{ (gas)}, \qquad (2)$$
$$\text{Al (liquid)} + H_2O \text{ (gas)} = AlO \text{ (gas)} + H_2 \text{ (gas)}. \qquad (3)$$

According to Webb [3], the AlO molecules are the main component of the gaseous phase, since its vapor pressure at 1500°C is 10^{-2} atm. When reaction (1) is the main reaction, then reaction (3) is the dominant secondary reaction. The water vapor reacts with liquid aluminum, forming $AlO_2 \cdot AlO$ which, being a metastable oxide, condenses on the solid Al_2O_3 and decomposes according to the reaction: $3AlO \text{ (gas)} = 3AlO \text{ (surface oxide)} = Al_2O_3 \text{ (whiskers)} + Al \text{ (liquid)}$.

However, Hargreaves [4] ascertained that the partial pressure of AlO in humid hydrogen is much lower than that of Al_2O at 1500°C, and therefore the main oxide in the gaseous phase is Al_2O. We find Hargreaves' argument very convincing.

The dislocational mechanism of growth has been proved for sapphire whiskers. The presence of a screw dislocation along the axis of the crystal has been confirmed by several investigations. Dragsdorf [5] found from the equatorial slope of Laue spots an elastic twisting of the crystal lattice in sapphire whiskers, while Eshelby [6] showed that this twisting is due to the presence of axial screw dislocations. Microscopic twisting of sapphire plates was observed with an ordinary optical microscope in [7]. The presence of a central capillary through the length of the whisker (Fig. 7) also indicates the presence of a screw dislocation. The dislocations in sapphire must have a large elastic energy concentrated essentially in the nuclei of the dislocations. This elastic energy exceeds the binding energy of atoms in the crystal lattice and leads to their disruption, and consequently to the formation of a capillary along the center of the crystal. We assume that sapphire whiskers grow by the mechanism of a screw dislocation when the material is added from the top.

LITERATURE CITED

1. P. L. Edwards and R. J. Happel, J. Appl. Phys. 33 (3): 826, 1962.
2. S. S. Brenner, J. Appl. Phys. 28: 1023, 1957.
3. W. W. Webb and W. D. Forgeng, J. Appl. Phys. 28 (12): 1449, 1957.
4. C. M. Hargreaves, J. Appl. Phys. 32 (5): 936, 1961.
5. R. D. Dragsdorf and W. W. Webb, J. Appl. Phys. 29 (5): 817, 1958.
6. J. D. Eshelby, J. Appl. Phys. 24: 176, 1953.
7. G. W. Sears and R. C. De Vries, J. Chem. Phys. 32: 93, 1960.

THERMOMECHANICAL TREATMENT OF TITANIUM ALLOYS
WITH A β-STRUCTURE

S. Z. Bokshtein, S. G. Glazunov, T. A. Emel'yanova,
Yu. N. Kabanov, S. T. Kishkin, and L. M. Mirskii

In recent years thermomechanical treatment of steels has been used extensively at home and abroad to obtain superhigh strength.

The purpose of this investigation was to determine the effect of thermomechanical treatment on two different titanium alloys; the compositions of the alloys are given in Table 1.

Thermomechanical treatment consisted in the following operations.

For the VT15 alloy: heating in the β-state at 760°C for 1 h; quenching in water from this temperature; aging at 450°C for 25 and 50 h.

For the V-120 alloy: heating in the β-state at 760°C for 1 h; quenching from this temperature in water; aging at 450°C for 16 h.

The results of short tests of samples treated in this manner are summarized in Table 2.

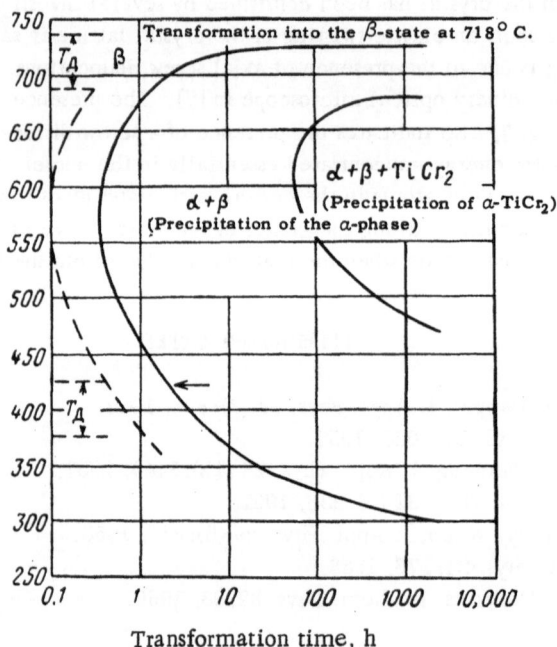

Fig. 1. Isothermal transformation curves of the V-120
alloy (Ti+13% V, 11% Cr, 4% Al).

TABLE 1

Alloy	Ti	Al	Mo	Cr	V	Impurities				
						Si	C	H	O	N
VT15	Base metal	3.76	7.80	10.7	–	0.15	0.10	0.015	0.15	0.01
V–120*	Base metal	3.1	–	11.6	12.60	0.13	0.11	0.015	0.14	0.06

* Paul D. Frost, Metal Progress, 1958, Vol. 75, No. 3, pp. 95-98.

TABLE 2

Alloy	Aging time at 450°C, h	Properties at 20°C				Properties at 400°C			
		σ_B, kg/mm²	$\sigma_{0.2}$, kg/mm²	δ, %	ψ, %	a_N, kg·m/cm²	σ_B, kg/mm²	δ, %	ψ, %
VT15	25	126	123	7.8	31.2	–	118	6.0	28
VT15	50	134	128	6.2	16.7	–	122	6.0	35
V–120	16	135	132	4.8	25	7.8	126	6.0	35

TABLE 3

Alloy	Aging time, h	Deformation, %	Properties at 20°C					Properties at 400°C				Strength after prolonged testing with $\sigma_B=100$ kg/mm²		
			σ_B, kg/mm²	$\sigma_{0.2}$, kg/mm²	δ, %	ψ, %	a_N, kg·m/cm²	a_N, kg·m/cm²	σ_B, kg/mm²	δ, %	ψ, %	Time before rupture, h	δ, %	ψ, %
VT15	25	10	153	146	3.0	11.3	1.7	127	5.2	31.5	–	–	–
VT15	50	10	147	141	2.6	7.6	1.2	117	5.0	31.5	–	–	–
VT15	25	45	159	155	3.0	10.6	1.1	123	6.0	38.2	13.5	19	49
VT15	50	45	152	149	4.2	12.1	1.3	Ruptured at the clamp			15	17.2	51.5
V–120	16	10	160	154	3.2	12.6	1.2	135	5.0	34	97	–	–
V–120	16	45	156	149	4.4	10	2.7	–	Ruptured at the clamp		Test interrupted after 100 h		

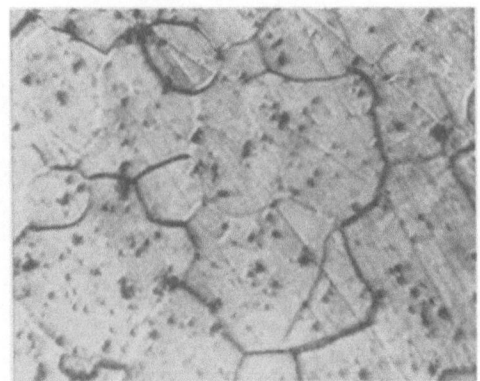

Fig. 2. Microstructure of the V-120 alloy in the initial state (quenched) in water from 760°C + aging at 450°C for 16 h). ×250.

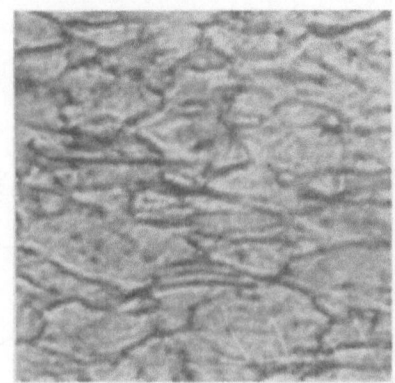

Fig. 3. Microstructure of the V-120 alloy after thermomechanical treatment (45% deformation+quenching in water from 760°C+aging at 450°C for 16 h. ×250.

The table shows that the mechanical properties of both alloys are rather good, and therefore even a small increase in the strength would be a considerable improvement (the high specific strength of titanium alloys should be taken into account).

The diagram of the isothermal transformation of the V-120 alloy (Fig. 1) indicates that there are two zones of stability of the β-phase in this alloy: in the supercritical temperature region and also below the bend in the C-shaped curve, the length of time being sufficient for plastic deformation to occur in the single phase region of the β-solid solution. This temperature range is arbitrarily indicated as T_d.

The presence of molybdenum instead of vanadium in the VT15 alloy may result in some change in the isothermal transformation curves. In this investigation the deformation temperatures were the same in both alloys: 760 and 350°C.

Deformation at 760°C. As was indicated earlier, both alloys were heated at 760°C in a muffle furnace for 30 min before being subjected to deformation and then rolled (10 and 45% swaging); they were then quenched immediately.

The samples were then aged at 450°C, the VT15 alloy for 25 h and 50 h and the V-120 alloy for 16 h.

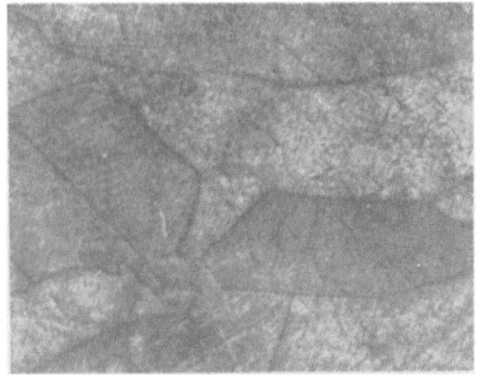

Fig. 4. Microstructure of the VT15 alloy after thermomechanical treatment (45% deformation + quenching in water+ aging at 450°C for 50 h). ×250.

Fig. 5. Microstructure of the VT15 alloy after thermomechanical treatment (deformation at 760°C, cooling in water to 350°C, 45% deformation quenching in water+ aging at 450°C for 50 h. ×250.

TABLE 4

Alloy	Aging time, h	Deformation, %	Properties at 20°C					Properties at 400°C		
			σ_B, kg/mm²	$\sigma_{0.2}$, kg/mm²	δ, %	ψ, %	a_N, kg·m/cm²	σ_B, kg/mm²	δ, %	ψ, %
VT15	25	45	160	155	3.1	23	1.0	124.5	3.5	21.2
VT15	50	45	154	148	2.9	11.0	1.1	122	4.0	23.8

The results of tests after aging are given in Table 3.

The comparison of the results obtained after aging and before aging of the VT15 and V-120 alloys (see Table 2) indicates that thermomechanical treatment increases the strength of the alloys and simultaneously decreases their ductility at room temperature, but increases the ductility at 400°C.

Deformation at 350°C. As in the preceding case, the samples were heated at the temperature of the β-state and then cooled in air to 350°C. They were then placed in a furnace at 350°C, where they were kept for 2-3 min, and then rolled (10 and 45% swaging). The sheets were quenched immediately after rolling.

It should be noted that we could not roll the V-120 alloy at 350°C because it split into layers parallel to the rolling plane.

The VT15 alloy was swaged 45% at 350°C in one or several passes.

Table 4 shows the properties of VT15 alloy after 45% deformation at 350°C followed by aging at 450°C for 25 and 50 h.

The table shows that 45% deformation of the VT15 alloy at 350°C followed by quenching and aging increases the strength, but decreases the ductility at room temperature, and increases the reduction of the section at 400°C after aging for 50 h.

The comparison of the microstructures of the V-120 alloys after ordinary treatment (Fig. 2) and after thermomechanical treatment (Fig. 3) shows that in the second case the grains become smaller, as in the case of steels.

No traces of recrystallization were observed after deformation and quenching (Figs. 3, 4, and 5).

On the micrographs of the VT15 alloy (Figs. 4 and 5) one can see dark-etched grains consisting of a mixture of elongated light and dark crystals. In the dark grains the orientation of slip lines (occurring as the result of plastic deformation) is probably more favorable for slip and nucleation of a new phase than in the light crystals.

The comparison of the results of short tests of the VT15 alloy with the same degree of deformation (45%) but different deformation temperatures indicates that the strength of this alloy after thermomechanical treatment at 350°C (Table 4) is somewhat higher than after thermomechanical treatment at 760°C (Table 3). At the same time, the relative elongation remains the same in both cases, and the reduction in section is somewhat greater at 350°C than at 760°C.

These preliminary results indicate that thermomechanical treatment strengthens titanium alloys with a β-structure.

The ultimate strength and the yield strength of the VT15 and V-120 alloys after thermomechanical treatment increase 18-20 kg/mm² as compared to the strength of the alloys treated in the ordinary way. The ductility at room temperature decreases somewhat, but is restored at 400°C.

The nature of the strengthening effect of the thermomechanical treatment is not yet clear.

EFFECT OF THERMOMECHANICAL TREATMENT
ON DIFFUSIONAL MOBILITY

S. Z. Bokshtein, S. T. Kishkin, and L. M. Moroz

One of the most promising new methods of increasing the strength of metals is thermomechanical treatment, i.e., plastic deformation combined with heat treatment.

In many cases thermomechanical treatment is used to improve the resistance of metals to high temperatures. We investigated the character of the diffusional mobility in alloys subjected to thermomechanical treatment as compared to that in alloys subjected to ordinary heat treatment. In what follows we describe the results of this investigation.

The alloys investigated were the ÉI437B alloy and ÉI481 steel. The alloys were tested after the following treatments: a) after ordinary heat treatment; b) after thermomechanical treatment consisting of 28% deformation at 1080°C at the rate of 13.5 m/min.

The diffusion rate was measured by the radioactive isotope method (diffusion in samples cut at an angle to the surface coated with radioactive material) and also by microstructural analysis and the absorption method.

The radioactive isotope method makes it possible to measure diffusional mobility through the grains and along the grain boundaries of the alloy. The average diffusion coefficients were determined by the absorption method.

The samples in their initial states and after thermomechanical treatment were coated electrolytically with Fe^{59}. (Before coating, the samples were subjected to electrolytic polishing to remove the cold-worked layer.) Then the samples were subjected to diffusional annealing in a vacuum furnace at 800°C for 150 h (a lower annealing temperature would have required a prohibitively long heating time). After diffusional annealing the diffusion coefficients through the grain and along the grain boundaries in the original state and after thermomechanical treatment were calculated. The results are summarized in Table 1. The average diffusion coefficients were calculated by the absorption method (Table 2).

TABLE 1. Diffusion Coefficients of Iron, $D \cdot 10^{13}$ cm^2/sec, Along the Boundary and Through the Grain of the ÉI437B and ÉI481 Alloys at 800°C

Alloy	Initial state		Thermomechanical treatment	
	D_g	D_b	D_g	D_b
ÉI481	0.62	4.6	2.8	–
ÉI437B	0.87	3.5	1.7	11

TABLE 2. Average Diffusion Coefficients $D \cdot 10^{13}$ cm^2/sec of Iron at 800°C

Alloy	Initial state	Thermomechanical treatment
	D_{av}	D_{av}
ÉI481	1.4	3.0
ÉI437B	1.0	1.3

Comparison of the results showed that the average diffusion coefficient in the alloys after thermomechanical treatment is greater than after ordinary heat treatment. This effect is particularly clear in the iron alloy. The results further show that after thermomechanical treatment the diffusional mobility increases through the grain as well as along the grain boundaries.

In the ÉI437B alloy the diffusion coefficient of iron at 800°C is doubled in the grain and tripled along the grain boundaries by thermomechanical treatment. In ÉI481 steel the coefficient of self-diffusion of iron through the grain at 800°C is quadrupled by thermomechanical treatment and the diffusion rate along the boundaries becomes impossible to determine because of the general diffusion through the whole volume of the crystal.

The increase in the diffusion rate is apparently due to the change in the fine structure of the ÉI437B and ÉI481 alloys as the result of thermomechanical treatment and also to the change in the state of the grain boundaries as the result of plastic deformation.

Figure 1 shows the autoradiogram and micrograph of the same area of the ÉI437B alloy (in the initial state) after the diffusion of radioactive iron into the alloy.

The autoradiogram shows that after ordinary heat treatment iron diffuses essentially along the grain boundaries in the ÉI437B alloy. After thermomechanical treatment the character of diffusion changes drastically. Diffusion becomes very intense along the slip lines and along the twins within the grains (Fig. 2a, b). Figure 3 shows that iron diffuses within the grain along the slip lines.

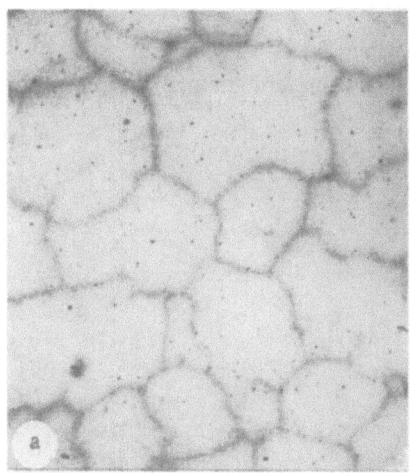 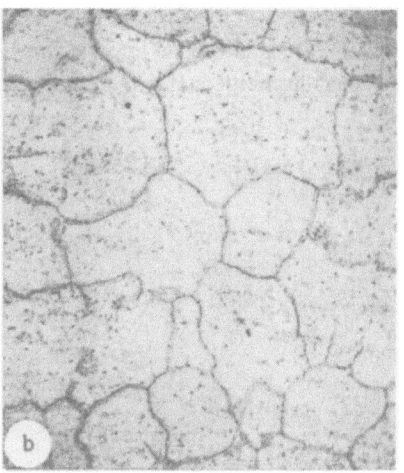

Fig. 1. Diffusion of iron in the ÉI437B alloy after ordinary heat treatment. ×50. a) Autoradiogram; b) micrograph.

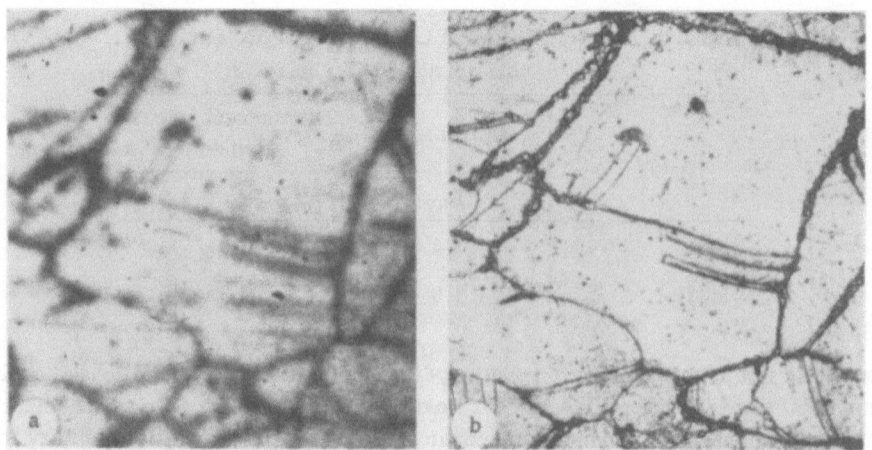

Fig. 2. Diffusion of iron in the ÉI437B alloy after themomechanical treatment. ×50. a) Autoradiogram; b) micrograph.

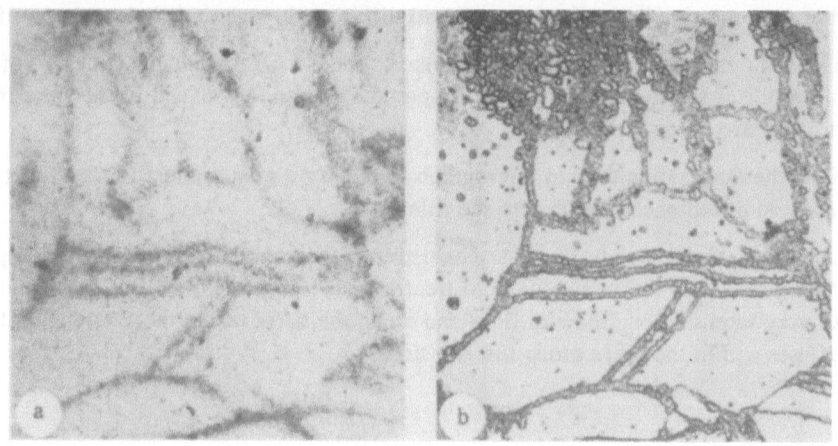

Fig. 3. Diffusion of iron through the grain along the slip lines in the ÉI437B alloy after thermomechanical treatment. ×50. a) Autoradiogram; b) micrograph.

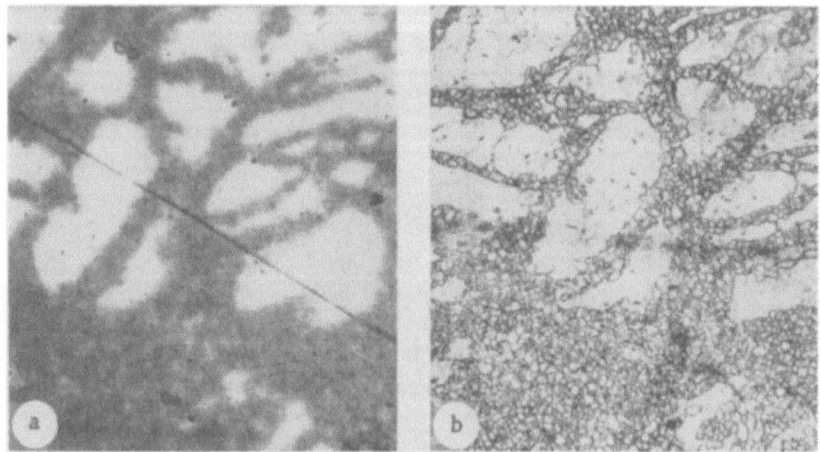

Fig. 4. Diffusion of iron in the ÉI437B alloy containing grains of different sizes after thermomechanical treatment. ×50. a) Autoradiogram; b) micrograph.

In the ÉI437B alloy thermomechanical treatment under conditions inducing partial recrystallization (deformation temperature 1080°C, 28% deformation, deformation rate 13.5 m/min) leads to the formation of a structure consisting of grains of different sizes; there are large areas of very small grains and also large grains bordered with small grains (Fig. 4a, b). As the result, the total length of the boundaries increases, and therefore the participation of boundaries in the total diffusional flow also increases. The same situation occurs in the ÉI481 alloy.

Thus, the increase of the mobility within the grains is probably due to the disruption of the regular structure of the metals as the result of plastic deformation, the formation of slip lines, and the development of substructure. As to the increase in the mobility along the grain boundaries, it is apparently due to the increase of the total length of the boundary because thermomechanical treatment induces a particular shape of boundaries and also because of the increase in the diffusional flow in the areas directly adjacent to the grain boundaries, since thermomechanical treatment destroys the crystal lattice in these areas.

The measurements of the diffusion rates through the grain and along the grain boundaries by the autoradiographic method indicate that thermomechanical treatment changes the fine structure of the crystals.

Consequently, thermomechanical treatment changes not only the state of the grain boundaries but also the state of the grains, and these changes are rather stable, since the diffusional effect remains even after prolonged annealing (of the order of 100 h or more). These results are in agreement with the results obtained in [1], where the authors showed that the increased diffusional mobility induced by plastic deformation remains even after annealing at a temperature considerably higher than the recrystallization temperature.

The increase in the diffusional mobility resulting from thermomechanical treatment, taking into account the present theory of the role of diffusion in heat resistance, limits the possibilities of applying thermomechanical treatment to heat resistant alloys. Apparently, thermomechanical treatment can be applied to alloys used at relatively low temperatures. At the same time, the increase in the diffusional mobility resulting from thermomechanical treatment can be useful because it increases the amount of the strengthening phase.

LITERATURE CITED

1. Effect of annealing on diffusion following plastic deformation, this volume, p. 40.